Tropical Botanic Gardens: Their Role in Conservation and Development

Edited by

V.H. Heywood and P.S. Wyse Jackson
Botanic Gardens Conservation Secretariat

ACADEMIC PRESS
Harcourt Brace Jovanovich, Publishers
London San Diego New York
Boston Sydney Tokyo Toronto

ACADEMIC PRESS LIMITED
24-28 Oval Road
London NW1 7DX

United States Edition published by
ACADEMIC PRESS INC
San Diego, CA 92101

**British Library Cataloguing in Publication Data
is available**

Printed in Great Britain at the University Press, Cambridge

This book is printed on acid-free paper
ISBN 0—12—346850—7

Contributors

Affolter, J. University of California Botanical Garden, Centennial Drive, Berkeley, California 94720, U.S.A.

Auge, P. L'Association des Responsables et Techniciens de Jardins Botaniques, c/o Institut National de la Recherche Agronomique, Villa Thuret, B.P. 2078 -06606 Antibes CEDEX, France

Bidin, A. Fernarium Universiti Kebangsaan Malaysia, Department of Botany, Faculty of Life Sciences, Universiti Kebangsaan Malaysia, 43600 Bangi, Selangor, Malaysia

Bramwell, D. Jardín Botánico Viera y Clavijo del Excmo. Cabildo Insular de Gran Canaria, Apartado 14, Tafira Alta, Las Palmas de Gran Canaria, Spain

Cuadros V., H. Jardín Botánico Guillermo Pineres, Apartado Aéreo 5456, Cartagena, Colombia

Cruttwell, N.E.G. Lipizauga Botanical Sanctuary, Eastern Highlands Provincial Government, P.O. Box 348, Goroka Papua New Guinea

Ducatillion, C. Jardin Botanique de la Villa Thuret, Institut National de la Recherche Agronomique, Villa Thuret, B.P. 2078 -06606 Antibes CEDEX, France

Dulloo, M.E. Forestry Department Curepipe, Mauritius

Forero, E. Missouri Botanical Garden, St Louis, Missouri, U.S.A.

Friedmann, F. ORSTOM, Muséum national d'Histoire naturelle, Laboratoire de Phanérogamie, 16 rue de Buffon, F 75005, Paris, France

Galland, J.P. Le Secrétariat d'Etat à l'Environnement, 14 Boulevard du General Leclerc, 92524 Neuilly-sur-Seine, France

Gómez Mejía, A. Jardín Botánico del Quindío, Apartado Aéreo 123, Armenia, Colombia

Hamann, O. Botanic Garden, University of Copenhagen, Oster Farimagsgade 2 B. DK-1353 Copenhagen K, Denmark

Heywood, V.H. Botanic Gardens Conservation Secretariat, Descanso House, 199 Kew Road, Richmond, Surrey, TW9 3BW, U.K.

Irawati. Herbarium Bogoriense, Research and Development Center in Biology, Lipi, Bogor, Indonesia

Lamb, A. Fruit Tree Division and Tenom Orchid Centre, Agricultural Research Station, P.O.Box 197, Tenom, Sabah, Malaysia

Lefebvre, B. Technical Counsellor Responsible for Fauna and Flora, Office of the Minister for the Environment, France

Lesouef, J.-Y. Conservatoire Botanique de Brest, Vallon du Stang-Alar, 52 Allée du Bot, 29250 Brest, France

Martinelli, G. Jardim Botânico do Rio de Janeiro, Rua Jardim Botânico 1008, 22.460 Rio de Janeiro, Brazil

Meilleur, B.A. Amy B.H. Greenwell Ethnobotanical Garden, Bernice Pauahi, Bishop Museum, P.O. Box 1053, Captain Cook, HI 96704, U.S.A.

Meredith, L.D. Australian National Botanic Gardens, Canberra, Australia

Mitchell, R.B. c/o The Cottage, Church Road, Coddenham, Ipswich, IP6 9PY, U.K.

Moutou, F. Laboratoire Central de Recherches Vétérinaires BP 67, F-94703 Maisons-Alfort, Cedex - France

Owadally, A.W. Forestry Department, Curepipe, Mauritius

Peixoto, A. Luna, Universidade Federal Rural do Rio de Janeiro, Jardim Botânico, Caixa Postal 74 582 23 851, Seropèdica, Itaguai RJ - Brasil

Randrianosolo, V. Department of Natural History, Parc Botanique et Zoologique de Tsimbazaza, BP 561 Antananarivo 101, Madagascar

Rasmussen, F.N. University of Copenhagen, Botanical Laboratory, 140 Gothersgade, 1123 Copenhagen K, Denmark

Rasmussen, H.N. University of Copenhagen, Botanical Laboratory, 140 Gothersgade, 1123 Copenhagen K, Denmark

Richardson, M.M. Australian National Botanic Gardens, Canberra, Australia

Riswan, S. Herbarium Bogoriense, Research and Development Center in Biology, Lipi, Bogor, Indonesia

Sastre, C. Muséum national d'Histoire naturelle, Laboratoire de Phanérogamie, 16 rue de Buffon, F 75005, Paris, France

Stewart, J. Sainsbury Orchid Fellow, Royal Botanic Gardens, Kew, Richmond, Surrey, TW9 3AB, U.K.

Strahm, W. c/o Forestry Quarters, Black River, Mauritius

Sukendar. Herbarium Bogoriense, Research and Development Center in Biology, Lipi, Bogor, Indonesia

Sumithraarachchi, D.B. Royal Botanic Gardens, Peradeniya, Sri Lanka

Van Vliet, G. CITES Secretariat, 6 Rue de Maupas, PB 78, 1000 Lausanne 9, Switzerland

Walter, K.S. The Center for Plant Conservation, 125 The Arborway, Jamaica Plain, MA 02130, U.S.A.

Wyse Jackson, P.S. Botanic Gardens Conservation Secretariat, Descanso House, 199 Kew Road, Richmond, Surrey TW9 3BW, U.K.

Programme committee

Professor P. ASHTON - President, International Association of Botanic Gardens

Dr J. BOSSER - O.R.S.T.O.M., Paris

Dr M. CHAUVET - Bureau des Ressources Génétiques, Paris

Mr D. FALK - Director, the Center for Plant Conservation, St Louis

Dr F. FRIEDMANN - O.R.S.T.O.M., Paris

Monsieur J.P. GALLAND - Secrétariat d'Etat à l'Environnement, Paris

Professor V.H. HEYWOOD - Director, Botanic Gardens Conservation Secretariat

Dr C. SASTRE - Muséum National d'Histoire Naturelle, Paris

D.B. SUMITHRAARACHCHI - Director, Royal Botanic Gardens, Peradeniya, Sri Lanka

Monsieur P. VALCK - Conservatoire et Jardins Botaniques de Nancy

Dr P.S. WYSE JACKSON - Botanic Gardens Conservation Secretariat

CONTENTS

Introduction 1

Address to the 2nd Congress on behalf of M. Brice Lalonde, Minister for the Environment, France
Bertrand Lefebvre 5

Part 1. Tropical Botanic Gardens: international issues

Developing a strategy for germplasm conservation in botanic gardens
Vernon H. Heywood 11

Needs and opportunities of South American botanic gardens for conservation
Enrique Forero 25

Towards an Australian Botanic Gardens Conservation Secretariat
L.D. Meredith and M.M. Richardson 35

International co-operation helping to promote botanic garden staff training in the tropics
Peter S. Wyse Jackson 45

French policy on the protection of wild flora, notably in the overseas territories
J.P. Galland 59

The work of WWF-France with regard to botanic gardens in the West Indies
Claude Sastre 71

Part 2. Tropical Botanic Gardens: case studies

The ethnobotanical garden and tropical plant conservation
Brien A. Meilleur 79

The Botanic Garden of Quindío: five thermic zones of Andean flora
Alberto Gómez Mejía 89

Measures that are required to help conserve the flora of Mauritius and Rodrigues in *ex situ* collections
A.W. Owadally, M.E. Dulloo & Wendy Strahm 95

The development of a university botanic garden in Brazil
Ariane Luna Peixoto 119

Rôle of botanic gardens in Sri Lanka and the conservation of medicinal plants
D.B. Sumithraarachchi 127

Cultivation on the French Mediterranean coast (the Côte d'Azur) of rare or threatened tropical or sub-tropical trees, two examples: *Jubaea chilensis* and *Cupressus dupreziana*
Pierre Auge & Catherine Ducatillion 151

The potential rôle of the Rio de Janeiro Botanical Garden in conserving the Atlantic Forest of Brazil
Gustavo Martinelli 163

Part 3. Tropical Botanic Gardens: their rôle in in situ conservation and reintroduction

Indigenous and alien plants in the Galápagos Islands: problems of conservation and development
Ole Hamann 169

The threatened plants of the flora of the Seychelles and their conservation
F. Friedmann 193

Botanic gardens in conservation: reintroduction into the wild
David Bramwell 209

From rescue to re-introduction; the example of *Ruizia cordata*
Jean Yves Lesouef 217

The rôle of the Fernarium as a sanctuary for the conservation of threatened and rare ferns, with particular reference to Malaysia
Aziz Bidin 223

The Lipizauga Botanical Sanctuary: a conservation project in Papua New Guinea
N.E.G. Cruttwell 241

Part 4. Plant Trade and CITES

CITES: Plant trade and botanic gardens
Ger Van Vliet 249

TRAFFIC-France against the traffic of tropical plant species
C. Sastre & F. Moutou 259

Part 5. Botanic Gardens and Orchid Conservation

Orchids in botanic gardens - necessity or luxury?
Joyce Stewart 267

The rôle of botanical gardens in orchid research
H.N. Rasmussen & F.N. Rasmussen 275

Orchid propagation in botanic gardens
Robert B. Mitchell 289

Orchid conservation in Bogor Botanic Gardens and its associated gardens
S. Riswan, Irawati & Sukendar 297

The conservation of orchids in Sabah, Malaysian Borneo
A. Lamb 307

Part 6. Short Communications

The potential for the development of botanic gardens in Madagascar and *ex situ* conservation of Malagasy threatened endemics
Voara Randrianosolo 331

The potential for the development of twinning and liaison with botanic gardens in the tropics: a western hemisphere example
James Affolter 333

Computerized plant record systems for botanic gardens
Kerry S. Walter 335

The Cartagena Botanic Garden, Colombia: an experience in the problems and possibilities of financing a tropical botanic garden
Hermes Cuadros V. 337

Resolutions of the 2nd International Botanic Gardens Conservation Congress
Resolutions Committee 341

Index 371

Introduction

In November 1985 the First International Botanic Gardens Conservation Congress was held in Las Palmas, Canary Islands, under the theme of "Botanic Gardens and the World Conservation Strategy" and as part of the Joint IUCN-WWF Plants Conservation Programme. This conference recommended that a Botanic Gardens Conservation Congress meet every three years.

Following a recommendation of the Las Palmas meeting, IUCN established the Botanic Gardens Conservation Secretariat at the beginning of 1987. The aim of the Secretariat is to help to co-ordinate and monitor the plant conservation activities of the botanic gardens of the world. The Secretariat provides technical guidance, information and support to enable gardens to meet the conservation challenges that are presented. It helps to build plant conservation and public education programmes in existing gardens and promote the establishment of new botanic gardens, particularly in the developing world, where often the greatest amount of species diversity is being lost.

The BGCS was established with the support of WWF and with grants from the Jessie Smith Noyes Foundation and the W. Alton Jones Foundation. It is a membership organization and is largely funded by the subscriptions of the member botanic gardens throughout the world. In January 1990 the BGCS became an independent organisation registered as a charity in the U.K.

At the Las Palmas Congress the involvement of Botanic Gardens in implementing the World Conservation Strategy was reviewed. A draft of the Botanic Gardens Conservation Strategy, prepared by Professor V.H. Heywood was presented to the Congress and discussed. Following extensive revisions this was presented in its final draft form to the Second Congress in Réunion and published in December 1989.

The Second International Botanic Gardens Conservation Congress was held in the Ile de la Réunion from 24 to 28 April 1989 under the theme "Tropical Botanic Gardens: conservation and development". It was attended by over 200 delegates from many countries and provided an opportunity to review the continued progress and involvement of botanic gardens world-wide in implementing the World Conservation Strategy. The particular problems of establishing, developing and strengthening botanic gardens and their programmes in the tropics was a major theme. Other topics included how best to foster regional networks and national programmes of botanic gardens, how to make CITES work better for plants and the rôle botanic gardens can play in the establishment of germplasm banks for wild species. A programme of papers, most of which are included in this book, and discussions was held in five sessions: 1. botanic gardens in the tropics, 2. regional networks and national programmes in the tropics, 3. germplasm banks in botanic gardens, 4. CITES and plants and 5. orchids: techniques in conservation and propagation.

The Congress was organized by the IUCN Botanic Gardens Conservation Secretariat, hosted by the Conservatoire et Jardin Botanique de Mascarin and held with the generous support of the Conseil Général de la Réunion and the co-operation of the French Government. Local arrangements were undertaken by M. Dominique Valck and his staff at the Conservatoire et Jardin Botanique de Mascarin.

This Congress was extremely successful and will help to direct and focus world attention on the needs for germplasm conservation in botanic gardens world-wide, especially in the tropics, and work towards the implementation of the Botanic Gardens Conservation Strategy. In his introduction to the Congress IUCN's Chief Scientist Plant Conservation and Director of the BGCS, Professor Vernon Heywood pointed out that "Botanic Gardens are globally concerned with conservation of the enormously rich biological resources of the tropics and seeking ways of developing and using them in a sustainable manner so that they will be available for future generations". In a presentation to the Congress he outlined the development of the BGCS and its programme and the recent advances made in germplasm conservation, particularly with regard to botanic garden activities and interests. He emphasized the need for greater integration of off site and on site conservation by botanic gardens, a view which was strongly endorsed by other speakers. He also stressed the need to expand existing botanic garden germplasm conservation facilities and establish many more new initiatives, especially for the preservation of germplasm of tropical crops, their land races and their wild relatives, as well as species of ornamental or scientific interest.

Throughout the Congress, speakers highlighted the new emphasis in the botanic garden community on the close scientific management of botanic garden germplasm collections and their increased concern with research and maintenance of *ex situ* plant genetic diversity.

At the concluding session of the Congress a series of resolutions was proposed and carried, put to the Congress by the Resolutions Committee, under the chairmanship of Dr Enrique Forero, Director of Research at the Missouri Botanic Garden, St Louis, U.S.A. The resolutions, stress the commitment of the participants to the aims and implementation of the Botanic Gardens Conservation Strategy, the need for close co-operation and extensive support of the Botanic Gardens Conservation Secretariat by the botanic gardens of the world and the further development of the botanic garden rôle in the conservation of tropical germplasm and the biodiversity of the tropics. In November 1989 the resolutions of the Congress were published in English, French and Spanish as a booklet. This book represents the published proceedings of the Réunion Congress.

The Third International Botanic Gardens Conservation Congress is due to be held in Rio de Janeiro, Brazil, in October 1992.

P. S. Wyse Jackson

---

1. French and Spanish translations of some of the papers in this volume were undertaken by Lesley Taylor and Dr Begoña Aguirre Hudson respectively, to whom the editors are very grateful.

2. Funding from the World Wide Fund for Nature (WWF) has enabled the editors to send copies of this book to botanic gardens in some developing countries.

Address to the 2nd Congress

M. Bertrand Lefebvre

Technical Counsellor Responsible for Fauna and Flora in the Office of Brice Lalonde, French Minister for the Environment

Protocol demands that in the absence of a Minister the Préfet who represents the state in any given department, speaks in his name.

However, before giving way to Monsieur le Préfet, with his agreement, I want to transmit a message of friendship to you from Brice Lalonde, the Secretary of State responsible for the Environment.

The Secretary of State, who had been hoping until yesterday that he would be able to come here today, wants very much to tell you that he would have been sincerely pleased to participate in the work of the Congress, and he would have been here wearing several hats. The first of these would have been a personal one, because Brice Lalonde is fascinated by botany and has belonged, as a very active member, to numerous associations for the protection of nature. Before now he has already had the opportunity to study the Réunion flora in the field. Secondly, Brice Lalonde, Secretary of State responsible for the Environment wishes to send you a double message. He wants to say to you that over and above the great projects which are being developed currently, and which are so publicised by the media (waste disposal, atmospheric pollution, the ozone layer, the control of natural and industrial hazards), the problem of the conservation of our animal and plant heritage is also one of our greatest priorities.

The second message is one, which the Minister wishes to address to the whole of the people of Réunion, that he would like to remind them that he follows, with great interest, all the problems in Réunion for which he is responsible, concerning the protection of genetic heritage, as well as the protection of habitats and landscapes, the prevention of hazards, coastal improvement and water management.

TROPICAL BOTANIC GARDENS: THEIR ROLE IN CONVERVATION AND DEVELOPMENT
ISBN 0—12—346850—7

Brice Lalonde wishes to remind Professor Heywood of France's commitment to the global programme for the protection of flora, carried out within IUCN under his authority. IUCN, of whose aims we have been reminded already, was created in 1949 in France at Fontainebleau, and the organization celebrated its 40th birthday in October last year in Paris and in Fontainebleau. On the occasion of this birthday, in the company of all its members - some 60 states and 450 associations - IUCN adopted the Fontainbleau Declaration, as it is called, which will promotes a non-aggression pact with nature, a pact which gives Man the right to his environment, but also responsibilities towards his planet.

As far as the protection of flora is concerned, France has chosen to implement a policy of global and dynamic protection, the effect of which is to afford a complimentary protection of spaces and species.

A specialist from the Direction Française de la Protection de la Nature will later on be developing this theme of France's conservation policy. Let us say for the moment that with regard to spaces, different protective measures have already been implemented: the classification of nature reserves (there is already one here in Réunion at Saint Philippe), or the constitution of biological forest reserves (6 have been created in Réunion thanks to the Office National des Forêts, covering some 7,000 hectares). With regard to the protection of species in their environment, different legislative measures also exist in France, the strongest being the integral protection one; I refer to a law on the protection of nature enacted on the 10 July 1976. In this regard 61 species have already been protected in Réunion Island.

But sometimes, alas, and in spite of the implementation of these different measures, species get very close to extinction and the rescue of remaining examples has to be implemented *ex situ*, often requiring the most sophisticated reproduction techniques. It is one of the rôles of our gardens and our botanic conservatories which are today revolutionizing conservation practices. These establishments store for us the genetic diversity of our plant world, feeding the development of our research in agricultural forestry, pharmaceutical and industrial domains, and represent a veritable economic stake in the race to save our natural heritage. Our conservatories are notably trying to achieve three goals: to gain information on and an understanding of selected plant species, the practice of conservation and the collection of data. They work with a wide network of qualified botanists and their concrete activities reveal a level of expertise which is very special and a unique approach in Europe. One can cite two current examples of this: the reintroduction on the island of Réunion of *Ruizia cordata* "bois de senteur blanc", of which only three plants remained in the wild, and which have been successfully cultivated by artificial crossing in the greenhouses of the Conservatoire Botanique de Brest, and 500 plants have been produced, which were recently replanted in the wild in their original environment in December 1988. One could also cite the conservation in the Conservatoire et Jardin Botanique de la Réunion of *Ochrosia borbonica*, "bois jaune", a rare species, which provide a product called ellipstisine, which is used in the treatment of certain cancers.

At a time when the disappearance of tropical forest appears to be a new threat for the planet, it would be difficult to end without mentioning the Réunion forest. Its flora is extremely diverse, and original to the island, containing numerous endemic species; it is now being listed and is part of our global heritage.

The Réunion flora has already been the subject of important protection measures. Some people think that the work should go further. An IUCN scientific mission has thoroughly examined the problem and has just published a report which we will study with great interest. Nevertheless, well before the coming into being of this mission, the Ministers of Agriculture, of Forests and of the Environment, as well as ONF, got together in recent months with the aim of looking at further possibilities for implementing even more protection. It is too soon today to be able to announce those measures which have been envisaged, but it does seem essential that from now on that a more dynamic policy of protection be implemented on the basis of a common initiative, a progressive policy and one to which everybody is sensitive: the state, scientists, forest specialists, associations, and so on.

Finally, and to conclude, I would like to underline the interest of the Minister in the Conservatoire et Jardin Botanique de Mascarin project. Let us all hope that the project grows rapidly, thanks to the enthusiasm of its founders, as well as that of the Conseil Général, which has entirely supported it. Let us hope that it will become a scientific cornerstone, as have its great metropolitan brothers in Brest, Porquerolles and Nancy.

Thank you very much.

Part 1

Tropical Botanic Gardens: International Issues

Developing a Strategy for Germplasm Conservation in Botanic Gardens

Vernon H. Heywood

Botanic Gardens Conservation Secretariat, Richmond, Surrey, U.K.

Summary

Botanic gardens have long been involved in the introduction and dissemination of plant resources, especially during the European colonial periods in the 19th and early 20th centuries. Today plant genetic resource conservation is largely undertaken by bodies such as FAO, IBPGR and crop gene banks around the world but these deal mainly with crop cultivars and only to a small degree with wild species, especially crop relatives. Botanic gardens are now the main centres for the conservation of wild species. Their concern is largely with rare and endangered species and are increasingly involved in programmes of their conservation and reintroduction into the wild. Many of the conservation collections in botanic gardens are, however, inadequate in terms of initial sampling, number of accessions and individuals, and form of maintenance. There is a need for botanic gardens to follow much more rigorous standards if they are to play an effective part in overall national conservation policies and genetic resource strategies. Cooperation between botanic gardens is also necessary as well as between gardens and national and private conservation agencies to ensure that integrated approaches to conservation are effective. Botanic gardens will have to play a leading advocacy role to develop financial support and wider public recognition for genetic conservation of plant resources on the massive scale that will be needed in the face of worldwide threats to biodiversity.

TROPICAL BOTANIC GARDENS: THEIR ROLE IN CONSERVATION AND DEVELOPMENT
ISBN 0–12–346850–7

Introduction

Although botanic gardens have been involved in conserving germplasm in one form or another for centuries, it is only in recent years that the concept of genetic resource conservation as such has developed through the work of bodies such as IBPGR, FAO and the CGIAR Crop Centres and the 150 or so germplasm collections and genebanks around the world. Most of the efforts of these bodies have been concentrated on a relatively small number of crop plants and their relatives, forage and pasture species and a selection of forest tree species. Not only have massive germplasm collections been built up for many of these selected species but a great deal of experience has been acquired of sampling, storage, germination and management techniques for the conservation of seed in seedbanks and for the maintenance of clonal collections and field genebanks.

Botanic gardens have a long tradition of introducing plants into cultivation and maintaining them for long periods, especially in the case of trees. They have been responsible for the introduction of many economically important plants into agriculture, horticulture and forestry and this continues to the present day, although much plant introduction is now carried out by specialist agencies. Many tropical botanic gardens were originally established as nurseries or introduction centres for germplasm of species that might prove useful either for the local economy or for the sponsoring European power.

The history of the role of botanic gardens in helping establish the agricultural economies of many parts of the tropics and subtropics has been well documented (Brockway, 1979 ; Holttum, 1970; Heywood, 1983; Purseglove, 1959 ; Smith, 1986) and the parts played by botanic gardens such as Bogor, Howrah (Calcutta), Pamplemousses and Singapore in introducing and developing plantation crops such as tea, coffee, oil palm, rubber and various spices are fully recognized.

In addition, botanic gardens in some temperate countries such as Australia and South Africa, which were considered by the early settlers to be devoid of plants of agricultural value, became involved in the introduction and selection of exotic plant materials for local use in agriculture and forestry. The Adelaide Botanic Garden, for example, during the directorship of Schomburgk (1865-1891) added an agronomic role to its educational and recreational functions. It introduced strains of wheat, oats and sorghum and tested fruits and vines which, if considered successful and suitable, were distributed to growers and the public. Likewise, botanic gardens in the Cape region of South Africa, from the days of Jan van Riebeck and the early Dutch settlers have been heavily involved in the introduction, assessment and dissemination of plants from various parts of the world. These ranged from the 50 or so crop plants introduced in the first few years of settlement at the Cape to the various trees and woody species, particularily from Australia, that were introduced through the Company Garden, by Baron von Ludwig through his own botanic garden, or by Cape Town Botanic Garden and used among other purposes for the stabilization of the Cape Flats. In North America, a continent relatively poor in plant genetic resources, the early botanic gardens developed by the European settlers, such as the Trustees Garden of Georgia at Savannah founded in 1733 and the first botanic garden in the United States founded by John Bartram in 1730 near Philadelphia and the Linnean Botanic Garden at Flushing, Long Island in the same year, were involved in plant introductions and relied largely on the experience of British gardeners and collectors.

Another area in which botanic gardens became involved was acclimatization of plants from other regions of the world. A number of acclimatization gardens were established such as the still extant Jardin de Aclimatacion de la Orotava in Tenerife, founded in 1788 with the aim of acclimatizing plants from the Spanish colonies in the New World before introducing them into cultivation in mainland Spain. In some parts of the world "Acclimatization Societies" flourished, such as that of Adelaide which used the large area north of the Botanic Garden known as the Botanic Park for its trials.

In the first half of the 20th century the economic role of botanic gardens diminished greatly. There were a few exceptions such as the Botanic Garden and Experimental Station of Lancetilla, Honduras, originally founded by the United Fruit Company as a plant introduction garden in 1926 to test new banana clones for disease resistance and to develop other new crops that might prove suitable for cultivation in the new world tropics. Amongst its achievements was the development of the banana variety "Valery" which was largely instrumental in reestablishing the banana industry of Central America. Today it maintains large collections of germplasm of tropical fruits including citrus and other Asiatic fruits and is currently being redeveloped.

The current situation

There are about 1500 botanic gardens and arboreta in operation in the world today. Not surprisingly they vary enormously in their involvement in conservation activities, including the preservation of plant genetic resources. Over 500 of them maintain conservation collections to some degree, mainly of native plant species, according to a survey undertaken during the preparation of the Botanic Garden Conservation Strategy (WWF/IUCN-BGCS 1989). In addition, the massive collections acquired and grown by botanic gardens today, for whatever purpose, represent the largest assemblage of plant diversity outside nature. There is a need, however, to consider carefully just what roles botanic gardens can play effectively in the conservation of plant genetic resources as there is a great deal of misconception about what is currently being done and about what potential there is.

It is important to recognize the differences between the conservation of germplasm for plant breeding as undertaken by the various crop gene banks and the conservation of samples of species as part of an overall strategy for their survival and reintroduction in the wild. In the former case, most of the germplasm accessions refer to cultivars of crop species and only a tiny percentage of the accessions (typically 2% according to Plucknett et al., 1987) are of wild species, usually crop relatives. Rigorous sampling techniques which have been developed over the years are employed in collecting the seed material of crop plants and most of the accessions are maintained as seed in banks under conditions of refrigeration and desiccation. However, in the case of species with recalcitrant seeds the material is kept as *ex situ* field genebanks, as clonal collections, in tissue culture or in a very few cases under cryopreservation. Most of the accessions in seed banks consist of several thousand seeds. Characterization and evaluation of the material is an important part of the process conservation of germplasm of crop cultivars although it has to be admitted that this is often not achieved in practice for some collections. Conservation of primitive cultivars (landraces) *in situ* is seldom attempted nor is *in situ* conservation of wild relatives a common procedure although crop breeders are becoming increasingly interested in the latter approach.

In the case of botanic gardens, germplasm conservation of major crop species as practised by the agricultural gene banks is not a primary concern although some botanic gardens do maintain conservation collections of germplasm of particular groups such as the tropical fruit trees, already mentioned, at Lancetilla, the IBPGR-recognized citrus collection at the Rimba Ilmu Universiti, Kuala Lumpur, Malaya, the taro (*Colocasia* spp.) and other economic collections at the Lyon Arboretum, Honolulu and the palm species and banana cultivars at the National Tropical Garden, Kauai, and the fruit tree germplasm collection at the Cibinong satellite garden of the Kebun Raya, Bogor.

Botanic gardens, especially in areas rich in plant diversity such as the Mediterranean and the tropics, can almost certainly be expected to play a much a greater part in this kind of germplasm conservation in the future although the emphasis will be more on wild species, including crop relatives, of economic importance such as under-exploited fruits, vegetables, rootcrops, oils, fibre plants and medicinal plants, rather than on cultivars of highly domesticated species. This will require careful planning and coordination and close cooperation with agencies such as IBPGR and FAO who have acquired extensive experience in genetic conservation over the past 20-30 years.

Most current conservation work by botanic gardens is focussed on rare and endangered species, either of the native flora or of exotic species belonging to groups in which the garden or associated institution has a particular interest. In Australia, for example about one third of the country's rare or threatened plants are in cultivation in botanic gardens, although half of these are in a single collection or even represented by a single plant (Meredith and Richardson, this volume) and it is hoped that a more comprehensive approach will be possible in the future. A similar situation obtains in South Africa where several gardens focus on the cultivation of native flora, including endangered species. North American examples include the Rancho Santa Ana Botanic Garden, Claremont, California, the Florida Rare Plants Project at Bok Tower Gardens, Lake Wales, Florida, and the Crosby Arboretum, Picayune, Mississippi. Many important conservation programmes on native flora are based at such gardens in North America, Europe and in other temperate countries.

In the tropics, usually only a minority of the native flora is in cultivation in botanic gardens in each country, largely because of the small number of gardens available and the lack of know how about the cultivation and propagation requirements of many of the species and the sheer difficulties of growing and maintaining many of them, but again several gardens are undertaking important research into the conservation and propagation of locally endangered species, sometimes with the aim of reintroducing them into the wild. Sometimes this work is done in association with other gardens in temperate countries in Europe and North America and several examples are mentioned in this volume such as the Programa Mata Atlantica at the Rio de Janeiro Botanic Garden.

It has to be recognized, however, that there is little consistency in approach or methodology in such conservation efforts and one of the greatest challenges facing botanic gardens is the promulgation and acceptance of appropriate standards and methodologies for all the procedures involved in long-term genetic conservation. I use the term genetic conservation deliberately since neither botanic gardens nor any other institutions can conserve whole species *ex situ* and it does a disservice both to the gardens and to the cause of conservation to suggest otherwise. We must resist the temptation to talk loosely about particular gardens conserving a certain number of species and saving them from extinction when what is meant is that the gardens have collected and are maintaining seed samples (not even necessarily under conditions of desiccation and refrigeration) or are growing a number of individuals, in the garden or *in vitro*, taken from some populations of the species concerned.

It is important, therefore, to be clear about the various approaches and goals in conservation that can be undertaken by botanic gardens. If the aim is indeed to save a species from extinction, then a fully integrated approach needs to be adopted as advocated by Falk (1990). This involves a detailed study of the field situation so as to establish the extent, status and viability of the remaining population(s) of the species, the nature of the threat(s) to its continued existence, consultation with the relevant conservation agencies or individuals concerned with the area, deciding on the most appropriate conservation strategy, working out costs and a timescale for implementation. The appropriate strategy may involve some or all of the following:

- collection of seed samples (following agreed guidelines), ramets or in certain circumstances whole plants;
- preservation of seed samples under conditions of desiccation and refrigeration;
- bulking up of seed samples;

- research into seed germination and reproductive biology;
- research into the cultural requirements of the plants;
- propagation and multiplication of the ramets;
- cultivation of material *in vitro*;
- raising and growing plants in a field gene bank or conservation plots;
- bulking up of material for reintroduction into the wild;
- reintroduction trials;
- monitoring of reintroduction experiments.

As Wallace and McMahan (1988) aptly write, "Bringing plants into cultivation is only the first step towards saving them from extinction. The second step is ensuring their survival in suitable protected sites in the wild." Of course, the number of species for which such rescue efforts can be undertaken is small considering the labour and costs involved. Very careful selection of priority species has to be made and the procedures adopted by the twenty plus botanic gardens working with the Center for Plant Conservation for the conservation of the most threatened plant species of North America are to be commended.

Certainly at a national level some form of coordination of effort between botanic gardens is needed. Ideally this should form part of a national germplasm strategy but in the absence of one, botanic gardens should organize their own national policy in association with the appropriate government and private agencies and prepare a national germplasm collections database which is essential for effective coordination of different efforts. This is much easier to achieve in countries such as the United States or in Europe where there are large numbers of botanic gardens potentially available to share out the work. It is encouraging to learn of the proposals by Australian botanic gardens to establish a national Botanic Gardens Conservation Secretariat (see Meredith & Richardson, this volume) and of moves to build botanic garden networks in several Latin American countries which will help to coordinate their conservation activities. Such networking has obvious financial benefits as will be discussed below.

While full scale conservation and reintroduction programmes are not possible for many species, individual botanic gardens can successfully and usefully engage in one or more of the activities listed above. For example, the collection and cultivation or storage of seed samples of rare and endangered species as a form of insurance policy is a valuable function for botanic gardens as is discussed below.

As regards what are often termed "living collections", the commonest form in which conservation materials are held by botanic gardens, a first priority must be to assess what is held by each garden. Globally the figure of species in cultivation in botanic gardens must run into tens of thousands, perhaps not far short of 100,000, as far as can be judged from the numbers of accessions that individual botanic gardens claim to hold. It is, in practice, exceedingly difficult in most cases to ascertain with any degree of accuracy just how many species are included in the total accessions of a botanic garden. Most of the totals cited for accessions do not distinguish between species, subspecies, varieties and cultivars, nor between different accessions of the same taxon. Claims by any garden to hold as many as 30,000 species in a collection must be treated with suspicion. Nonetheless, several gardens do appear to have over 10,000 species represented in their collections and a 5,000 total is quite common. In addition, many gardens have built up specialist collections of impressive size. For example, important orchid collections include those of the Royal Botanic Gardens, Kew (c.3,000 species), the Orchid Jungle, Homestead, Florida (2,000 species and 15-20,000 hybrids), the Wheeler Orchid Collection and Species Bank at Bell State University (3,000 species) and the Smithsonian National Orchid Collection, Washington DC (2,500 species) while many tropical botanic gardens have smaller but significant collections of native orchid species, maintained both *in situ* in sanctuaries and *ex situ* in the garden. Examples include, in India, the Botanic Garden, Orchid Research and Development Centre, Tippi, Arunachal Pradesh which grows 400 native species, the National Orchidarium and Botanic Garden, Shillong with a collection of 200 native species, and the National Orchidarium and Experimental garden, Yercaud, with 210 native species and there are similar native orchid gardens or collections in other tropical countries such as Malaysia (see paper by Lamb, this volume).

Because of the importance of orchids and the very large number of orchid collections worldwide, the Botanic Gardens Conservation Secretariat is planning to undertake a complete survey and add the data to the BGCS Collections Database so that an up-to-date assessment can be made available of the state of orchid conservation collections. This is part of the policy of BGCS in making surveys of conservations collections in botanic gardens for major taxa, such as bromeliads, carnivorous plants, palms, conifers, or for geographical areas such as China, Cuba and the Indian Ocean islands.

It has long been customary for botanic gardens to publish catalogues of their holdings, although many gardens have discontinued them, either for economic reasons or in view of the rapid turnover of material which soon makes the catalogues out of date. In recent years. a number of gardens have revived their catalogues such as the National Botanic Garden of Belgium at Meise which has published a catalogue of its greenhouse collections, containing 17,000 accessions, a substantial number of which are of wild origin. A Catalogue of Succulents in Dutch Botanic Gardens has recently been published the Dutch Botanic Garden Foundation (Groens 1990). From this it has emerged that in the 13 gardens surveyed, besides the Cactaceae, at least 34 out of the 51 families with succulents listed in Jacobsen's "Das Sukkulentenlexikon" are in cultivation. The publication of such catalogues has been made much easier by the computerization of records by many gardens. A major problem with such catalogues, however, apart from keeping them up-to-date, is ensuring the accuracy of the information given and the nomenclature used. This is a problem they share with Seed Lists (see below). Also there is such a high degree of overlap between the catalogues of different botanic gardens that is not possible to make an asessment of the total holdings represented globally. BGCS reviews a number of these catalogues critically to extract from them species to add to the Collections Database. When the catalogue of the Bogor Botanic Garden was screened, it was found that a considerable number of the 3,500 species listed were Indonesian native species not yet represented in the BGCS database.

If so-called living collections are to be taken seriously as germplasm or genetic resources, for short- or medium-term conservation, reintroduction, breeding, selection or other purposes, there are certain requisites that will have to be followed by botanic gardens. Apart from adequate initial sampling, as already mentioned, the size of the sample should be adequate for the purpose. It is clearly not acceptable to regard a single or a few individuals as an adequate conservation sample and wherever possible 10-50 individuals should be grown in the case of trees or shrubs and at least 50 in the case of perennials or biennials. This will seriously reduce the number of species that can be conserved in this way for reasons of space, finance and labour.

The location of conservation collections also needs consideration so that certain safeguards regarding the possibilities of crossing with other species are observed. While it is possible to grow some of these in the main public areas of the garden or arboretum, provided security can be maintained. Gardens should consider setting aside special areas away from public view for the cultivation and maintenance of conservation material in the same way that experimental stations grow material in rows in plots. Many gardens will not have space for such experimental plots and some alternative location or additional area may have to be sought. The problem is more acute for older urban botanic gardens where space is at a premium than for more recent gardens, especially those where only a part of the area allocated for the garden is actually landscaped. In planning a new botanic garden with a conservation policy provision should be made for special conservation plots and reserve areas, with associated propagation and other facilities to be included. Selections of conservation material make a valuable demonstration for public education and good examples include the display garden of local rare and endangered species at King's Park and Botanic Garden, Perth, W. Australia and the demonstration bed of endangered species at Cambridge University Botanic Garden, U.K.

There remains the question of what should be done about the many thousands of rare and endangered species which are in cultivation in botanic gardens in minimal samples of one or a few specimens, either in the open or under glass. Provided the material is accurately identified and of known wild origin, urgent steps should be taken to ensure the propagation of the plants so that the risks of its loss or of genetic deterioration that may occur in such small samples are minimized. Duplicate material should be distributed to other botanic gardens willing and able to maintain the collections, especially those in the native geographical or climatic range of the species concerned. Although the BGCS Collections Database is by no means comprehensive, the annual summaries of rare and endangered plants in cultivation it sends out to member gardens often reveal to a garden that it is the only one to hold a particular species and that steps should therefore be taken to ensure its survival.

Similar considerations apply to the many species which are today found in botanic garden collections only, having disappeared in the wild. While this is situation strongly to be deplored it is urgent for gardens which have such material to report the facts to BGCS, if they have not already done so and endeavour to propagate the material so that it can be distributed to other gardens willing to grow it under suitable conditions.

The collection of seed from the wild and from plants growing in the garden and the making this seed freely available to other gardens throughout the world through the Seed List or *Index Seminum* is a botanic garden tradition dating back over two centuries. Currently over 600 gardens produce such lists which vary enormously in their style, scope and contents. The whole future of the Seed List in the context of plant conservation will need review and some recommendations are included in the 'Guidelines for the *ex situ* conservation of germplasm by botanic gardens' prepared for BGCS by a working group (Bramwell & Heywood, 1991). In particular, with the emphasis placed on seed of wild origin for scientific and conservation work and on adequate sampling and storage, the bulk of the of the seed traditionally listed in botanic garden seed lists which is garden-collected from plants of unknown wild origin is of limited interest except for certain horticultural purposes. Those requesting seed from an *Index Seminum* should ask for data on the origins of the material to be received, unless it is already given, so as to encourage gardens to provide such information in their Seed Lists as a matter of routine. Some botanic gardens separate wild-collected seed from other seed in their lists, others no longer include seed of unknown wild provenance, while a few no longer issue a traditional seed list but produce a seed bank catalogue. A number of gardens are prepared to collect and supply seed of particular species on request.

As mentioned above, the collection of seed samples of locally endemic species from the wild, especially those that are rare or endangered, as a contribution to conservation is a valid procedure for botanic gardens, provided that it is recognized that rigorous procedures for sampling, storage and handling have to be followed if the material is to satisfy the strict criteria that apply to genetic resources. Obviously any seed of a rare or endangered species collected in the wild is of some value but for conservation purposes this is limited unless the requisite procedures are followed. Since the development and maintenance of a seed bank is an expensive operation, botanic gardens much explore the possibilities of using facilities at a national or regional crop seed bank. Alternatively botanic gardens within country could share a common botanic garden seed bank and contribute to its running costs. This underlines the need for countries to develop a strategic policy for the conservation of plant genetic resources in which botanic gardens are an essential component.

An important consideration in the development of a germplasm strategy, which is seldom adequately addressed, is the economic evaluation of the genetic resources which it is intended to conserve and more generally the valuation of the plant diversity from which they are sampled. Even in the case of crop genetic resources few attempts have been made to make specific analyses of the economic benefits, largely because it is now generally accepted that the importance of such genetic conservation for food security is self-evident. It has been estimated by Plucknett et al., (1987) that worldwide expenditure on germplasm amounts to no more than US$ 55 million which is a trivial sum in relation to what is spent on military security. Any assessment of the value of wild plant genetic resources in simple economic terms is bound to be highly speculative, except in a small number of cases, compared with major crop genetic resources where it is much easier to demonstrate their value to humankind. Advocacy must play a great part in obtaining support for wild plant genetic resource conservation on the massive scale that will be necessary in the face of the worldwide threats to plant diversity. Botanic gardens will have to be in the forefront of such efforts.

Reference has been made to the need for coordination of genetic resource conservation at national and regional levels. Action is needed also at the international level and BGCS is attempting, within available resources, to maintain an overview of conservation collections and to provide guidance for the formulation and implementation of a global strategy. Particularily important is the need for adequate documentation of germplasm collections in botanic gardens and the adoption of a standard methodology for this, using, for example, the *International Transfer Format for Botanic Garden Records* (ITF). This can also serve to conserve data on samples of germplasm moved between institutions.

References

Bramwell, D. & Heywood, V.H. (1991) *Guidelines for Germplasm Conservation by Botanic Gardens*. Botanic Gardens Conservation Secretariat, Richmond and Las Palmas de Gran Canaria.

Brockway, L.H. (1979). *Science and Colonial Development: the Role of the British Royal Botanic Gardens*. Academic Press, London and New York.

Groen, L.E. (1990). *Catalogue of Succulents in Dutch Botanic Gardens*. Compiled for the Dutch Botanic Gardens Foundation. Stichting Nederlandse Plantentuinen, Boskoop.

Heywood, V.H. (1983). Botanic gardens and taxonomy – their economic role. *Bull. Bot. Survey India,* 25, 134-147.

Holttum, R.E. (1970). The historical significance of botanic gardens in S.E. Asia. *Taxon,* 19, 707-714.

Purseglove, J.W. (1959). History and functions of botanic gardens with special reference to Singapore. *Gardens' Bulletin, Singapore*, 17(2), 125-154.

Plucknett, D.L., Smith, N.J.H., Williams, J.T. & Anishetty, N. Murthy (1987). *Gene Banks and the World's Food*. Princeton University Press, Princeton.

Smith, N. (1986). *Botanic Gardens and Germplasm Conservation*. University of Hawaii Press, Honolulu.

Wallace, S.R. & McMahan, L. (1988). A place in the sun for plants. *Garden*, 12(1), 20-22.

Needs and opportunities of South American botanic gardens for conservation

Enrique Forero

Missouri Botanical Garden
St Louis, Missouri, U.S.A.

Summary

Botanic gardens should play a very important rôle in the conservation of plant genetic resources. This can be accomplished through research, education, *ex situ* and *in situ* conservation programmes, and plant exploration and development. Opportunities to carry out these objectives are promising in South America. There are currently about 58 botanical gardens in the region. Although they work mostly in isolation, some efforts have been made to coordinate activities at the national level in several countries.

Botanic gardens in South America should organize themselves and decide on their priorities, objectives, goals and conservation strategies. To this end they might consider the establishment of a network of South American botanic gardens. However, a real commitment on the part of international coordinating bodies and funding agencies is urgently needed for such a network to become a reality.

Introduction

Of the total of approximately 250,000 species of plants, nearly 90,000 grow in the neotropical region. Just in the countries of Colombia, Ecuador and Peru the richness is estimated at between 45,000 and 60,000 species of flowering plants. Brazil has one of the largest floras in the world, estimated at ca. 50,000 to 55,000 species.

TROPICAL BOTANIC GARDENS: THEIR ROLE
IN CONVERVATION AND DEVELOPMENT
ISBN 0—12—346850—7

However, if the destruction of the natural vegetation continues at its current rate, as many as 60,000 species of plants will be lost during the next 30-40 years.

Deforestation depends upon a number of related factors, underlined by economic and social problems facing the developing countries of the world. Population growth, pressure on available forest resources, industrial development, road construction, poor planning and lack of environmental impact studies have contributed considerably to this process of destruction. Contradictory government policies and lack of communication between government agencies, the scientific community and non-governmental organizations has led to mismanagement and ill-informed conservation measures.

Many useful products such as foods, insecticides, pesticides, essential oils, spices, gums, etc., are derived from plants. More than 50% of modern medicines come from the natural world. Plants are being tested for anti-cancer and anti-AIDS compounds.

Even though plants have many uses, less than 1% of them have been screened for their potentially useful properties. In fact, people presently depend on as few as 20 plant species for over 85% of their food. As the natural vegetation is destroyed, plants suffer a decline in their genetic diversity, and our own and future generations are deprived of the opportunity to retain and expand the availability of food crops, medicines and many other useful products.

Botanic gardens should play a very important rôle in the conservation of plant genetic resources because of their involvement in research, training, *ex situ* and *in situ* conservation programmes, and plant exploration and development. Although the number of botanic gardens in South America is limited, there is great potential for expansion and cooperation.

The World Conservation Strategy and the Botanic Gardens Conservation Strategy give botanic gardens the framework to develop conservation initiatives.

The Botanic Gardens Conservation Strategy has been established to stimulate the involvement of botanic gardens in the implementation of the World Conservation Strategy. The strategy provides the rationale for such involvement and gives policy guidance on how this can be achieved.

Accordingly, botanic gardens should contribute to the preservation of plant genetic resources and should help ensure the sustainable use of plant species and the ecosystems in which they occur.

In order to achieve these goals, the Strategy suggests that botanic gardens make a commitment to conservation through mission statements and improved standards of management. Also considered in the Strategy are the relationships between wild and managed conservation, as well as the importance of education, community involvement and training with emphasis on conservation.

Botanic gardens ought to consider conservation as one of their most important objectives. A well developed botanic garden may be more important now than at the time when societies were mostly agricultural and, therefore, closely linked to nature. A modern botanic garden has to provide visitors with an outline of the plant kingdom, its laws and its uses, so that they can admire it and learn to respect nature (Mangenot & Valck, 1987: 51).

People need to learn more about the importance of plants as sources of useful products, about the serious threats of extinction to a large proportion of plants, and about the importance of preserving the world's genetic resources (Eloff, 1987: 99). A modern botanic garden should be able to integrate conservation and development at some level, through the development of agroforestry ecosystems, rural energy development, and the cultivation of wild plants of potential economic value (Xu Zaifu, 1987: 249-251).

If botanic gardens in South America are to be effective in implementing conservation related strategies, it is urgent for them to plan and carry out programmes that take into account the specific needs of the region.

These and other aspects of the current situation of South American botanic gardens will be analyzed in the course of this presentation, and specific suggestions will be made to strengthen these botanic gardens and their rôle in conservation.

Background

Interest in botanic gardens in tropical and subtropical South America has increased considerably during the last ten or fifteen years. In some countries, both governmental and non-governmental organizations have demonstrated interest in creating new botanic gardens or in supporting the development of those already in existence.

According to the latest account by the BGCS, there are 58 botanic gardens in South America. In addition to the 19th century Rio de Janeiro Botanic Garden in Brazil, several major new botanic gardens have been established in recent years throughout South America, and most of them are active in research, teaching, conservation and exploration.

Although these botanic gardens work mostly in isolation, some coordination of activities at the national level is already underway. For instance, a meeting of directors of Argentinian botanic gardens was held during the national botanical congress in that country in 1987. A network of botanic gardens has recently been established in Colombia. A meeting of directors of Brazilian botanic gardens was held in Fortaleza in January of 1990.

The plants conservation programme was established by IUCN/WWF to promote, formulate and support efforts to conserve plants in developing countries in the tropics, "thereby contributing to the preservation of a sustainable environment" (Hamann, 1989: 31).

Since botanic gardens are a key component of the strategy, and because no other institutions are as directly related to the conservation of plants, they should be made capable of addressing these issues effectively.

Botanic gardens have become important centres for popular education about conservation and the importance of plants. They can be instrumental in local efforts to modify school curricula in order to make them more appropriate to meet the current needs in the areas of conservation and proper management of natural resources.

Research carried out in botanic gardens in South America can be fundamental in monitoring the decline of plant diversity and also to the conservation of plant genetic resources, including wild plants of actual or potential economic value. Botanic gardens in those countries can be very important for both *in situ* and *ex situ* conservation programmes. They should take responsibility for the threatened and endangered plants of their regions, which may not be known in detail, but which will at least be then recognized in terms of their existence in badly damaged or threatened habitats.

Botanic gardens should become repositories of materials needed for research on endangered plants as it relates to their future conservation, through reintroduction into the wild or otherwise (Ashton, 1987: 117). The strategy for wild species, for example, may need to consider both *in situ* conservation where possible and *ex situ* in seed banks. Botanic gardens should become actively involved in seed-banking and tissue-culture, including research into the preservation of recalcitrant seeds, which may constitute about a third of the plants of tropical lowland forests. Hawkes (1987: 146) has suggested that for recalcitrant seeds of tropical timbers and fruit trees, *ex situ* plantations and meristem banks may be the only solution to the storage problem, although the amount of genetic diversity that can be conserved by these means is only a fraction of the amount possible for seed banks. Additional problems to be considered for the conservation of species in seed banks include low seed production in the field, or simply not enough individuals in the natural populations to give a sufficiently large sample for seed-banking, parasitization in seeds, etc. (Bramwell, 1987: 179). Of the 80,000 to 90,000 species of plants in Latin America, about 10,000 to 15,000 species may be orchids, which have special preservation requirements.

In general, *ex situ* collections should be properly documented, should maintain an adequate representation of genetic diversity, and should make material available for research, for the assessment of economic uses, and for reintroduction into the wild.

National and international networks

According to Heywood (1987: 17), even though a great deal of attention and expertise has been focused on genetic conservation of plant species and populations, the ultimate responsibility for the conservation of thousands of endangered wild species and of species that have ecological importance, scientific importance, or local economic potential rests with the botanical community. It is in addressing these issues that botanic gardens must play a major rôle.

It is of the utmost importance for botanic gardens to organize themselves and to decide on their priorities, goals, objectives, and conservation strategies. They will then be in a better position to provide leadership for the considerable number of organizations concerned with problems of the environment that have been established in most South American countries. The growing conservation movement in South America includes universities, NGOs, and a greater involvement of the local communities in general. In cooperation with these conservation groups, botanic gardens should be able to build the capacity to conserve the tropical flora and therefore expand their own rôle in conservation by promoting it at the local level.

As has already been indicated, some networking at the national level is underway, and some steps have been taken towards increased communication and collaboration with botanic gardens in other parts of the world, particularly in the United States and Europe. There exists a very capable community of scientists eager to make a contribution to conservation, and their knowledge and expertise should be taken into account. Local initiatives should be considered seriously, strengthened and multiplied. A concerted effort is needed to tap existing resources within the countries and to match them with external funds.

I suggest that, at the national level, the Center for Plant Conservation of the United States provides a good example of the network approach to conservation of rare and endangered species, an example that could be followed, with modification, in some South American countries. The Center relies heavily on existing organizations, information and infrastructure. This model may be adapted to the particular resources and conservation needs of other countries. The primary goal of the Center is to create a complete programme of *ex situ* conservation, including appropriate collection, storage, cultivation and research for the nearly 3,000 extremely rare, threatened or endangered native plant species of North America (Thibodeau & Falk, 1987: 285-286).

The National Association of Botanical Gardens of Mexico provides an example of the way in which networking has been approached in Latin American countries. The objectives of the association are:

1. To establish contact with other botanic gardens in the country and to exchange ideas and information by publishing a newsletter, by

organizing conferences, symposia, etc., and by forming a bank of botanical and bibliographical information.

2. To promote the establishment of other botanic gardens within the country, to form a representative collection of the local flora by studying different ecosystems and by suggesting possible sites for botanic gardens.
3. To contribute to preventing the extinction of native species and to promote the use of native flora.
4. To promote financing for research, especially for courses and for purchase of materials in bulk.
5. To promote botanic gardens and an understanding of their importance (Pattison, 1987: 347-348).

It is here suggested that, in order for botanic gardens in the developing countries of South America to organize themselves and to plan a major rôle in the preservation of the plants of the region, they might consider the establishment of a network of South American Botanical Gardens.

Conclusions

The Botanic Gardens Conservation Strategy has been established to stimulate the involvement of botanic gardens in the implementation of the World Conservation Strategy. Botanic gardens should contribute to the preservation of plant genetic resources and to help ensure the sustainable use of plant species and the ecosystems in which they occur. Of particular importance are the relationships between wild and managed conservation, as well as the rôles of education, community involvement and training with emphasis on conservation.

Research, education and conservation opportunities are promising in South America, especially considering the growing awareness on the part of the general public in all the countries involved with regard to the urgent environmental issues facing them today.

After a few years of intense activity and rapid development, South American botanic gardens are now, for the most part, simply struggling for survival. There is great potential for these botanic gardens to play an important rôle in conservation. However, the lack of positive action on the part of the international community to support botanic gardens both technically and financially has resulted in considerable frustration.

Although some progress has been made since the First International Botanic Gardens Conservation Congress in 1985, much more direct involvement and encouragement is urgently needed. It is possible to point to some specific actions and call them "progress," but they are just not enough.

Of particular importance is the emphasis on collaboration among botanic gardens at the national and regional levels. Some networking at the national level is already taking place, but additional efforts are needed. Steps have also been taken towards increasing communication and collaboration with botanic gardens in other parts of the world. Such efforts should be strengthened and multiplied.

The necessity for the formulation of a carefully designed network of botanic gardens both nationally and internationally supported throughout the region should be stressed. A network of South American botanic gardens would, in principle, be able to play a major rôle in preserving the plants of the region. These botanic gardens should take some kind of responsibility for the threatened and endangered plants of their regions, which might not be known in detail, but which will at least be recognized in terms of their existence in badly damaged or threatened habitats. Most people would regard this as the highest international function of the botanic gardens and the major reason for international support.

However, a real conceptual or programmatic commitment on the part of the BGCS and the IABG, particularly to obtain the support of international aid organizations, the World Bank, the International Development Bank, and other funding agencies for the establishment of a network, is urgently required. A concerted effort is also needed to tap existing resources within the countries and to match them with external funds.

References

Ashton, P.S. (1987). Biological considerations in in situ vs. ex situ plant conservation. In: *Botanic Gardens and the World Conservation Strategy*, eds. Bramwell, D. Hamann, O., Heywood, V. and Synge, H., pp. 117-130. Academic Press, London and New York..

Bramwell, D. (1987). The rôle of the Jardín Botánico Canario "Viera y Clavijo" in the conservation of endangered Canarian endemics. In: *Botanic Gardens and the World Conservation Strategy*, eds. Bramwell, D. et al., pp. 175-181.

Eloff, J.N. (1987). Botanic gardens and education in South Africa. In: *Botanic Gardens and the World Conservation Strategy*, eds. Bramwell, D. et al., pp. 85-100.

Hamann, O. (1987). The IUCN/WWF Plants Conservation Programme in action. In: *Botanic Gardens and the World Conservation Strategy*, eds. Bramwell, D. et al., pp. 31-43.

Hawkes, J.G. (1987). A strategy for seed banking in botanic gardens. In: *Botanic Gardens and the World Conservation Strategy*, eds. Bramwell, D. et al., pp. 131-149.

Heywood, V.H. (1987). The Changing rôle of the botanic garden. In: *Botanic Gardens and the World Conservation Strategy*, eds. Bramwell, D. et al., pp. 3-18.

Mangenot, F. & Valck, P. (1987). A botanic garden for our city and university and a conservatory: the example of Nancy. In: *Botanic Gardens and the World Conservation Strategy*, eds. Bramwell, D. et al., pp. 47-51.

Pattison, G. (1987). The Mexican Association of Botanic Gardens. In: *Botanic Gardens and the World Conservation Strategy*, eds. Bramwell, D. et al., pp. 347-348.

Thibodeau, F. & Falk, D.A. (1987). Building a national *ex situ* conservation network - the U.S. Center for Plant Conservation. In: *Botanic Gardens and the World Conservation Strategy*, eds. Bramwell, D. et al., pp. 285-294.

Xu Zaifu. (1987). The work of Xishuangbanna Tropical Botanical Garden in conserving the threatened plants of the Yunnan tropics. In: *Botanic Gardens and the World Conservation Strategy*, eds. Bramwell, D. et al., pp. 239-253.

Resumen

Los jardines botánicos deben jugar un papel muy importante en la conservación de recursos genéticos vegetales. Esto puede lograrse a través de la investigación, educación, programas de conservación *ex situ* e *in situ*, y búsqueda y aprovechamiento de especies vegetales. Hay muy buenas posibilidades de lograr estos objetivos en América del Sur, donde existen actualmente 58 jardines botánicos. Aunque estas instituciones trabajan por lo general en forma aislada, ya se han hecho algunos esfuerzos para coordinar sus actividades a nivel nacional en varios países.

Los jardines botánicos en América del Sur deben organizarse y decidir sobre sus prioridades, objetivos, metas y estrategias conservacionistas. Para lograrlo, podran considerar la creación de una Red Suramericana de Jardines Botánicos. Sin embargo, se requiere con urgencia que las entidades coordinadoras y financiadoras internacionales tomen la decisión de apoyar a este consorcio para que se convierta en realidad.

Towards an Australian Botanic Gardens Conservation Secretariat

L.D. Meredith and M.M. Richardson

Australian National Botanic Gardens
Canberra, Australia

Summary

It is estimated that there may be 20,000 to 25,000 species of vascular plants in the continent of Australia and its associated islands. Valuable studies into the Australian flora are proceeding with the "Flora of Australia" project providing a focus for much detailed taxonomic assessment.

Australia's national listing of plant species at risk, "Rare or Threatened Australian Plants", has recently been revised and includes a total of 3,387 species of vascular plants, this number being made up of 221 endangered species, 853 vulnerable species, 1,371 rare species and 811 species that are poorly known but suspected to belong to one of the other categories. 131 species are listed as "believed extinct in the wild".

Australia has a large number of parks and gardens which fulfil some of the functions of a "botanic garden" but only 5 are scientific institutions employing botanists or other scientists engaged in studies of the Australian flora. Several institutions other than botanic gardens are, however, also working on the native flora.

During 1987 a survey of Australia's botanic gardens was carried out to determine what species of rare or threatened Australian plants were being grown. - 1,053 species, or about one third of the total list, were found to be in cultivation, but 515 of those were in only one collection, sometimes represented by only a single plant. The survey identified defects in the collection of threatened species but also highlighted

TROPICAL BOTANIC GARDENS: THEIR ROLE IN CONVERVATION AND DEVELOPMENT
ISBN 0—12—346850—7

important conservation work being carried out in the country's botanic gardens.

The report on the survey called for more co-ordinated efforts by Australia's botanic gardens towards conservation. Recommendations were made for networking, regional co-operation and national *ex situ* conservation policies. It was recommended that a "secretariat" for the co-ordination of the conservation activities of Australian botanic gardens be established.

Australia is fortunate in having a comprehensive national list of its native plants at risk. It was first published in 1981 and continually updated with input from botanists and other specialists, many of them revising and writing for the "Flora of Australia" project.

In 1984 a Working Group on Endangered Flora of the (Australian) Council of Nature Conservation Ministers decided that assessment of the Australian flora at risk would be aided by knowledge of the extent of cultivation of rare or threatened species in botanic gardens and arboreta. The survey was delayed until a complete revision of the 1981 list was undertaken. By late 1986 that revision was well advanced and funds were provided by the Australian National Parks and Wildlife Service to undertake the survey and the Australian National Botanic Gardens was contracted as the consultant. In addition to assessing the holdings of rare or threatened plants in botanic gardens the consultancy called for recommendations on the rôle of botanic gardens in conservation.

Mindful of the various meetings held to discuss the rôle of botanic gardens in the implementation of the World Conservation Strategy a methodology for the survey was devised, with an intention to reinforce calls for an increased effort by the botanic gardens of the world towards plant conservation.

A report issued in 1984 by the Royal Australian Institute of Parks and Recreation identified 55 gardens, parks or arboreta which in the opinion of the authors of the report fulfilled at least some of the functions of a botanic garden, these being defined as "to collect, study, exchange and display plants for research, education and enjoyment". Several new gardens could now be added to that total. That report also introduced a numerical system of classification for gardens based on features such as having an associated herbarium, record systems, security of tenure and so on. That system was referred to in the Draft Botanic Gardens Conservation Strategy. So when we consider those gardens in the first category, that is those having a "collection incorporated with an associated herbarium, scientific and technical staff, adequate recording system, continuity of purpose and security of tenure assured" we find only four qualify. These are the Royal Botanic Gardens, Sydney, the Royal Botanic Gardens, Melbourne, the Botanic Gardens of Adelaide and the Australian National Botanic Gardens at Canberra. Grouped with them we would add a "class 2" garden, Kings Park and Botanic Garden in Perth, as, although it does not maintain a large herbarium, it does employ botanists and other scientists. All of these gardens are managed by either federal or state government bodies. Botanic gardens in the capital cities in Queensland, the Northern Territory and Tasmania do not have an associated herbarium or scientific staff at this stage but do have important rôles to play in *ex situ* conservation. Many other gardens away from the major cities are primarily public parks with emphasis on horticultural display. They are important cultural features in local and regional settings and they too could be brought into a national conservation network.

The compilers of the publication, "Rare or Threatened Australian Plants", which had undergone extensive revision and was close to completion at the time of the survey, made available a copy of the revised list so that we were able to work from the latest information. Based on an estimated total flora of 20,000 species, about 17% of that total is regarded as rare or threatened.

The revised list included 3,387 species of vascular plants comprising:

- 131 species believed extinct in the wild
- 221 endangered species
- 853 vulnerable species
- 1,371 regarded as rare, and
- 811 poorly known species.

Australia uses the category "K" to denote poorly known species that are suspected, but not definitely known, to belong to any one of the other categories.

The list included 496 undescribed species and there have been some slight changes to the list since the survey was carried out.

Twenty four collections of Australian plants were considered for the survey. These included all the major botanic gardens, a number of smaller gardens known or thought to hold collections of Australian plants and some significant private collections and research institutions. The survey was carried out during visits to the various gardens by searching records to determine plant holdings and by interviewing gardens' managers to establish attitudes towards conservation. When possible, in the case of endangered species, a physical inspection of the plants was carried out.

All of the collections surveyed had at least some rare or threatened plants in their collection and some had large numbers. A number of gardens had conservation policies but it was apparent that there had been little co-ordinated effort towards developing a conservation collection and many, perhaps most, plants had been "fortuitously" acquired. At the end of the survey, 1,053 species, or about one third of the total list, were found to be in cultivation. Almost half of these, however, were present in one collection only and in some cases represented by only one plant. This survey did not consider Australian plants which are cultivated in botanic gardens outside Australia nor exotic species of rare or threatened plants that are grown in Australian botanic gardens and the importance of such collections should be recognised.

Managers of the gardens visited were interviewed to try to assess their attitudes to conservation. Several gardens were heavily involved in conservation issues and some were carrying out important research, particularly on their local species of threatened plants. All gardens that undertook education programmes had used conservation as a theme, either directly in school-age education programmes or indirectly in more passive displays or in plant labelling or descriptive brochures.

All persons interviewed thought that conservation was a "good thing". Indeed it was like motherhood - everyone was for it. However it was clear that many institutions were working in ignorance of what others were doing. Some plant families were well represented in living collections whilst others were not represented at all. For instance, *Eucalyptus* or *Grevillea* is widely grown but no botanic gardens are growing the 7 species of *Carex* on the list of rare or threatened plants. This may reflect the various collectors' personal interests and indicates a need for a more co-ordinated policy.

As a result of the survey it was possible to make a number of recommendations and the Draft Conservation Strategy was used as a guide (Heywood, 1985). It was recommended that all Australian botanic gardens play an increasing rôle in the conservation of rare or threatened Australian plant species and that they implement the IUCN/WWF Botanic Gardens Conservation Strategy as part of the World Conservation Strategy.

Other recommendations concerned training, regional links (that is within the South Pacific and South-East Asian regions and within Australia itself), and collections policy, such as collections of locally threatened plants, national collections of genera and guidelines on field collecting and the dispersal of *ex situ* collections to other botanic gardens to help ensure the security of such collections.

The report also called for a national meeting of all concerned in the rôle of botanic gardens in the conservation of rare or threatened plant species and recommended that an Australian Botanic Gardens Secretariat be established to co-ordinate the work of Australian botanic gardens in conservation and to liaise with international agencies.

It was recommended that the proposed Secretariat be based at the Australian National Botanic Gardens. This was not empire building on the part of the authors! The Australian National Botanic Gardens is the only major botanic garden in Australia that specialises in the study and display of the Australian flora. It has played a leading rôle in research into the biology and cultivation of Australian plants and it has a demonstrated commitment to conservation. The institution has initiated or co-operated in several conservation-related projects which are leading to a greater understanding of the Australian and related floras. And the establishment of such a body at the Australian National Botanic Gardens was suggested by directors of several other botanic gardens - a vote of confidence?

There is much good work being done in Australian botanic gardens. Kings Park and Botanic Garden at Perth, Western Australia, is carrying out research into the biology and horticulture of the threatened flora of that state and at the time of the survey was establishing a display garden, accessible to the public, to show visitors some of Western Australia's threatened plants and educate them in a quiet way to the problems. The Botanic Garden of Adelaide carries out similar research into the flora of South Australia. In Victoria, the Royal Botanic Gardens at Melbourne is developing a written policy on the conservation of threatened species and co-operates with other organisations on conservation-related issues. The Royal Botanic Gardens, Sydney, has recently opened a large arboretum for the cultivation of native plants which will provide facilities for an increased emphasis on threatened species. The small North-Coast Regional Botanic Garden at Coffs Harbour, New South Wales, has a display garden where rare or endangered plants are grown, with labels outlining the reasons for the plant's scarcity or the threat faced. As mentioned earlier, the Australian National Botanic Gardens has developed a conservation programme over a number of years and all sections of the institution, living collections, herbarium, botanical, biological, and horticultural research units and public programme and education sections are involved.

The problems facing the smaller botanic gardens are acknowledged and understood. Most are operated by local government parks departments with many calls on the community purse and never enough money. But frequently these gardens are in busy provincial centres, in holiday resorts or are on tourist routes. With co-ordination and encouragement they can play a part in a national conservation effort. For instance, the major botanic gardens, when designing displays could make them portable so that they could be circulated to the smaller gardens, so assisting in the educational process. Staff at the smaller gardens could be invited to join field trips which are planned for their area and some of the plants subsequently grown could be distributed to those gardens to disperse the collection. Assistance may be given in plant identification and labelling and staff could spend time working at the major botanic gardens as part of a co-ordinated training scheme.

Perhaps most importantly, it should be noted that all of Australia's major botanic gardens are in the temperate region of the continent. It has been suggested that Australia reflects the world botanic garden imbalance but with the compass points reversed: botanic gardens in the temperate south with the largely poorly known tropical flora in the north where there are no major gardens. We need the knowledge of personnel in the regional gardens, and local growing conditions to help us to cultivate species for study.

The report was handed to the National Parks and Wildlife Service with the recommendation that it be published and made available to interested parties. The report incorporated the draft revised list of rare or threatened plants which at the time of the survey was expected to be published before the end of 1987. For a number of reasons that list has not yet been published and this in turn has held up the publication of our own report.

The title of this paper is "Towards an Australian Botanic Gardens Conservation Secretariat". We all know that progress is seldom as fast as we would wish. There is as yet no Australian Secretariat. But the survey that I have described was the first step. In October 1988 the Federal Minister for the Environment announced the establishment of an Endangered Species Advisory Committee " to formulate and co-ordinate a national strategy to conserve endangered species and their habitats." The Committee will work in conjunction with various Australian bodies, as well as with IUCN and WWF and is interested in the rôle of botanic gardens in plant species conservation.

I believe that we are on the way to achieving some of the recommendations of the report. I hope that at the Third Congress we will be able to tell you of further progress that we have made.

References

Leigh, J., Briggs, J. and Hartley, W. (1981) *Rare or Threatened Australian Plants*. Australian National Parks and Wildlife Service, Canberra

Royal Australian Institute of Parks and Recreation (1985) *Report on the Collection of Native Plants in Australian Botanic Gardens and Arboreta*, Canberra

Briggs, J. and Leigh, J. (1988) *Rare or Threatened Australian Plants*, Australian National Parks and Wildlife Service, Canberra

Meredith, L.D. and Richardson M. M. (1987 unpublished) *Rare or Threatened Australian Plant Species in Cultivation in Australia*, Australian National Parks and Wildlife Service, Canberra

Heywood, V. H. (1985) *Botanic Gardens Conservation Strategy-First Draft*, IUCN/WWF

Resumen

Hacia un Secretaría para la Conservación en los Jardines Botánicos de Australia.

En el continente australiano e islas asociadas se estima que existen aproximadamente 20.000 a 25.000 especies de plantas vasculares. El proyecto de la "Flora of Australia" continúa aportando valiosos estudios de flora, sirviendo además y en particular como foco para el estudio taxonómico.

Recientemente se ha revisado la lista nacional de especies con riesgo o en vías de extinción: "Rare or Threatened Australian Plants"(Plantas australianas raras o amenazadas). Esta lista incluye un total de 3.387 especies de plantas vasculares. La cifra incluye: 221 especies que corren peligro de extinción, 853 especies que son vulnerables, 1.371 denominadas raras, y 811 que no son bien conocidas, pero que se sospecha pertenecen a una u otra de las categorías anteriores. 131 especies figuran como "extintas" en la naturaleza.

Australia tiene numerosos parques y jardines que cumplen algunas de las funciones de un jardín botánico, pero solamente 5, son instituciones científicas que cuentan con botánicos u otros científicos que desempeñan estudios sobre la flora australiana. Existen además varias instituciones que no son jardines botánicos, pero que también están realizando trabajos sobre la flora nativa.

Durante 1987 se hizo un estudio sobre los jardines botánicos de Australia para determinar cuales de las especies raras y/o amenazadas se cultivan en dichas instituciones. 1.053 especies, o sea una tercera parte de la cifra total, se encontraron en cultivo, pero 515 de ellas pertenecían a una sola colección, a veces representada por un solo ejemplar. Por medio de este estudio, se identificaron defectos en la manera de recolección de las especies amenazadas, pero al mismo tiempo, se destacaron trabajos importantes sobre la conservación que se realizaban en los jardines botánicos del país.

Al concluirse el estudio, se hizo notar la necesidad de una mayor coordinación entre los jardines botánicos, en sus esfuerzos para la conservación de las especies. Se hicieron recomendaciones para la interacción, cooperación y un plan de acción nacional *ex situ*. Para lograr mejor estos fines, se recomendó la creación de una Secretaría de Conservación para Jardines Botánicos Australianos.

Resumé

Sur le continent d'Australie et ses îles, on compte environ 20,000 à 25,000 espèces de plantes vasculaires.

Des études importantes de la flore australienne sont en cours, des études relatives au travail qui se poursuit pour "Flora of Australia" et ce qui fournissent bien l'occasion d'une évaluation taxonomique très détaillée.

La liste australienne des plantes menacées "Rare and Threatened Australian Plants" a été revisée récemment et le total s'élève à 3,387 espèces de plantes vasculaires, ce chiffre comprennant 221 espèces menacées, 853 espèces vulnérables, 1,371 espèces rares et 811 espèces peu connues mais que nous croyons appartenant à une des autres catégories. 131 espèces sont sur la liste comme étant "estimées disparu dans la nature".

L'Australie possède un grand nombre de parcs et de jardins qui entreprennent certaines fonctions de jardin botanique, mais il n'y en a que 5 parmi eux qui servent d'institutions scientifiques et qui emploient des botanistes ou des scientifiques qui se préoccupent activement de la flore australienne. Plusieurs institutions en dehors des jardins botaniques se penchent, néanmoins, sur les études de la flore indigène.

Pendant l'année 1987 une étude a été entreprise pour déterminer quelles espèces de plantes rares ou menacées étaient en cultivation dans les jardins botaniques d'Australie. 1053 espèces, ou un tiers de la liste totale, ont été identifiées comme étant en cultivation, mais 515 de celles-ci se trouvaient dans une seule collection et étaient parfois représentées par une seule plante. Cette revue, tout en démontrant des faiblesses dans la collection nationale des espèces menacées, à servi également à souligner le travail important de conservation actuellement en cours dans les jardins botaniques du pays.

Le rapport qui a suivi l'étude a fait appel aux jardins botaniques australiens de coordiner leurs efforts de conservation. Des recommendations ont été portées en ce qui concerne un réseau de jardins botaniques, la coopération régionale et la politique nationale envers la conservation *ex situ*.

Il a été recommandé également qu'un Secrétariat soit établi pour la coordination de toutes les activités de conservation dans les jardins botaniques australiens.

International co-operation helping to promote botanic garden staff training in the tropics

Peter S. Wyse Jackson

Botanic Gardens Conservation Secretariat
Richmond, Surrey, U.K.

Summary

Promoting the development of botanic gardens in the tropics will be dependant on a number of factors, amongst which access to well trained and experienced staff resources must rank as one of the most important. Staff training in the fields of horticulture, conservation, botanic garden management and administration, public education and fund-raising will also be vitally important, as well as in the academic fields of botany, ecology, forestry, etc.

The paper explores the present state of international co-operation in training for botanic gardens staff in the tropics. It outlines some of the successes and failures in the past and points towards new initiatives presently being developed which may suggest strategies for progress in the future. Although some botanic gardens, especially in temperate regions, provide valuable training places for botanic garden staff in the tropics, this training is often not adequately geared towards the problems they will face in the future on returning to their gardens. As well as that, there are few if any courses available for many important aspects of botanic garden and plant collection management.

Many of these problems can be best solved or alleviated through widespread international, regional and national co-operation and by the development of training courses in many more botanic gardens than are presently in operation. Such courses can be initiated either by individual gardens or shared and organized by a network of botanic gardens and other institutions in a region.

TROPICAL BOTANIC GARDENS: THEIR ROLE
IN CONVERVATION AND DEVELOPMENT
ISBN 0–12–346850–7

The paper also focuses on some elements of the financing required for such developments and the rôle that the Botanic Gardens Conservation Secretariat hopes to play in this field.

Introduction

In the past there was much useful co-operation between botanic gardens in temperate and tropical regions. A considerable number of tropical gardens were founded under the aegis of the botanic gardens and other scientific institutes of such colonial powers as Britain, the Netherlands, France and Germany. Important tropical botanic gardens such as those at Bogor, Indonesia, Pamplemousses, Mauritius and Peradeniya, Sri Lanka owe their origins to such co-operation.

For much of this century close international co-operation between botanic gardens declined and often annual exchange of an *Index Seminum* was the only remnant of former close links. Following national independance there was an uneasy transition from colonial management to autonomous botanical institution in many tropical botanic gardens. Often the botanic garden managers of these great gardens were been trained in Europe and came to work in the gardens as expatriates. When this ended the organization of such gardens suffered considerable setbacks with a lack of well-trained local staff to take over. Some are only now beginning to recover. As well as that, many botanic gardens lost the purpose for which they were originally established; no longer required as gardens of introduction they became little more than public parks, albeit botanically very rich ones, with little emphasis on active scientific use of the collections. The rise of the global conservation movement has injected new life and purpose into the network of tropical botanic gardens, it is now up to the world botanic garden community to assist in their re-emergence.

Development of tropical botanic gardens

Promoting the development of botanic gardens in the tropics will be dependant on a number of factors, amongst which access to well trained and experienced staff resources must rank as one of the most important. Staff training in the fields of horticulture, conservation, botanic garden management and administration, public education and fund-raising will also be vitally important, as well as in the academic field of pure botany and forestry. We all know that a botanic garden with a well trained and motivated staff can become an important resource for conservation. Indeed, the importance of this is all too often illustrated by the virtual collapse of many botanic garden programmes when staff change or are reduced due to financial cut-backs. Many botanic garden managers should carefully consider the time when they will be replaced and pay more attention to preparing those that will follow, rather than lamenting the decline of a particular garden once they have moved on to new pastures. Perhaps we should reflect that such losses are often, at least in part, due to our own lack of foresight.

Throughout the world, botanic garden directors are in urgent need of assistance in their garden development programmes. One of the most frequent requests is for assistance with staff training, generally directed to the largest and best known botanic gardens, especially those in Europe and North America. These gardens have responded as best they can and many tropical botanic gardens have benefitted from their help and the new expertise and ideas that a well-trained staff member can give. Nevertheless these staff training places are few in number, often difficult to locate and can be expensive to provide. The small number of botanic gardens offering training courses are generally well aware of their limitations and urgently need their load to be shared. As well as that, many of the recipients of such training given in a botanic garden in the developed world benefit so much from their new skills that they are soon moved on to important administrative posts away from the botanic garden shortly after their return to their native countries.

International co-operation for training

Many examples of close co-operation between botanic gardens in the tropical and temperate regions are well known. The Royal Botanic Gardens, Kew, maintains a number of links with gardens in developing countries, most notably in recent years at Trivandrum in India and Limbe in Cameroon. Several students from the tropics spend training periods at Kew annually and their contribution towards training students from the developing countries has been neatly managed to become integrated with their general training course for horticultural students from the British Isles. Kew links and activities with other botanic gardens are outlined by Simmons et al. (1987), presented to the Las Palmas Conference in 1985 and will not be repeated here. Other examples elsewhere of co-operative staff training schemes have been developed in such gardens as Rancho Santa Ana in California which organizes a linking and staff exchange programme with botanic gardens in Mexico; in the U.K., the Northern Horticultural Society's garden at Harlow Carr provide a training place in horticulture for a student from Ecuador in 1988 as the first part of a co-operative programme which they hope to extend in the future (Swindells, 1987). The Bonn University Botanic Garden in West Germany has offered a one year training place for a staff member of a South American botanic garden (W. Lobin, pers. comm., 1988). The curator of the Lipizauga Botanical Sanctuary in the highlands of Papua New Guinea recently spent a year at the Royal Botanic Garden, Edinburgh (N. Cruttwell, pers. comm., 1989), (and their co-operative link has included sponsorship of the P.N.G. sanctuary as a member of the BGCS). Other examples of international co-operation are agreements between the National Botanic Gardens of Cuba and with the botanic gardens of Jena in Germany and Cordoba in Spain (Anonymous, 1988). A further one has been developed between the University of Berkeley Botanic Garden, California and the Robert and Catherine Wilson Botanic Garden in Costa Rica outlined by Affolter elsewhere in this volume.

Despite the great demand and need for training places, some botanic gardens receive few, if any requests for such assistance at present. Although many of them are small institutions with only limited resources and funding available to share, they often have a great willingness and ability to respond with genuine, valuable and generous help. Indeed, many small and relatively poverty-stricken botanic gardens in the developed world have risen to the conservation challenge magnificiently and with great resourcefulness and efficiency. Their staff often have the proven ability to pass on the skills most urgently required in a developing tropical garden where every penny must be turned before it is spent. These smaller gardens are a great resource for assisting tropical botanic garden development which is still almost untapped.

The rôle of the Botanic Gardens Conservation Secretariat

Since its establishment in 1987 the BGCS has been inundated with requests for aid especially from tropical botanic gardens and those in developing countries. These requests range from ones for general assistance with redevelopment and re-establishment of existing tropical botanic gardens to advice on the creation of completely new gardens. Such inquiries often include considerable staff training requirements. In many cases the BGCS has been able to respond positively to these appeals or to pass them on to colleagues in botanic gardens where assistance can be given. Nevertheless many of these queries have had to be acted upon in a less than satisfactory manner while we continue to develop the international co-operative structures described in this paper and required for the implementation of the Botanic Gardens Conservation Strategy (Heywood, 1989).

The main requests for assistance received from developing gardens are for 1. help with staff training, 2. guidance on policy for the garden, especially in the field of plant conservation, 3. building up of collections, 4. advice on record systems and 5. general financial help. The importance of the need for staff training facilities was also shown by responses to a questionnaire circulated to every known botanic garden as part of the development of the Botanic Gardens Conservation Strategy. In addition, these further indicated that there is great willingness on the part on the vast majority of the world's botanic gardens to enter into some form of co-operative arrangement with a garden in another part of the world and to receive or offer assistance with staff training. One aspect of such co-operative links are presently being developed as the BGCS gardens' "twinning" scheme. However, the greatest obstacle to the development of such links and co-operation in staff training was seen to be a shortage of finance.

Other difficulties for wider development of co-operative staff training programmes were seen by these gardens in their questionnaire responses to be presented by the administrative burden that organizing training courses presents, the difficulties of communicating over long distances, language, cultural and climatic differences, national political problems and "red tape". Nevertheless the opportunities that such co-operation and links can present for individuals and their gardens were generally agreed to out-weigh any of these drawbacks. Amongst the opportunities specificially mentioned were how such co-operation can help to reduce the isolation felt by staff of botanic gardens in inaccessible places or in countries with few other gardens; how important chances for staff career development are offered; how co-operation can lead to an increase in awareness of plant conservation needs, exchanges of ideas and, on a personal level, many new international friendships.

Financial considerations and training courses

Without doubt lack of finance has been, and will continue to be the greatest obstacle in the development of widespread staff training opportunities for tropical gardens. Nevertheless there are various ways which this problem may be reduced, if not entirely solved.

In most cases, realistically, the financial responsibility for staff training courses (at least if they are specifically designed for students and staff from developing countries) will have to rest with those gardens providing the training opportunity and be seen as one of their contributions towards plant conservation, rather than as a money-making scheme. Probably many gardens will have to view such courses as loss making, or at best break-even, services. Botanic gardens staff in the developed world are, in any case, often more likely to be effective in fund-raising and well versed in the skills required to seek out funding bodies, both national and international, as well as to prepare and present persuasive and attractive looking funding applications to such aid agencies. It has been suggested that in most cases the organizers of training courses should attempt to seek funding for the programme itself, rather than seeking sponsorship for each student. While applications for support for individuals may be successful for a year or two, such aid is rarely sustainable after the initial few years and early enthusiasm. Botanic gardens must become increasingly professional and imaginative in their funding approaches, especially if the considerable wealth of educational trusts, commerce and industry are to be successfully tapped and convinced of the importance of botanic gardens for plant resource conservation.

On the individual level, it is reasonable to expect that a person being trained should receive some support from his or her parent garden, and continue to receive a salary during their absence on study leave. Depending on individual family or home circumstances, this may considerably assist their participation in an overseas training course. International travel costs are often heavy but can be reduced by careful advance planning and through assistance from a variety of aid agencies and sources, such as the academic staff exchange programmes that operate in many countries and institutions. Many gardens have indicated to the BGCS that if a considerable element of practical work is included in training programmes then payment can be given. One European botanic garden suggested it might fill a staff vacancy on an annual basis with a trainee from a tropical country, for which a comprehensive practical training would be received by the candidate and a gardener's salary paid in return for their work in the garden, alongside the regular staff. Accommodation can often be provided cheaply in hostels and, for shorter training courses, in student accommodation in university or college botanic gardens and as paying guests in the homes of members of the garden's Society of Friends.

One should remember that it is not always necessary that a candidate need receive a very long period of training. Often an absence of one year in a temperate country places considerable strain on the trainees themselves and their families, not to mention the staff left behind who must continue to operate the essential functions of the botanic garden without the usual quota of staff. Short and intensive training courses are, in many cases, an ideal solution or at worst, a second best but workable alternative.

Another training course was proposed by a small university botanic garden in Europe recently which has been discussed with the BGCS. The garden suggested it would offer a training course annually for one or two candidates from developing world botanic gardens. The salary from one unfilled staff vacancy would be used to provide accommodation and some living expenses for the students. In return the trainees would spend half of their time working in the various departments of the garden, alongside the permanent staff and learning by a combination of actual supervised instruction and by example. The rest of their time would be spent sitting in at relevant undergraduate lectures in courses of ecology, plant physiology, taxonomy, etc. at the associated university botany department nearby. Study visits would also be made to co-operating botanic gardens, agricultural and horticultural research centres in the region where facilities for the commercial production of crops and ornamentals, disease control, crop breeding, seed banking and micropropagation programmes can be studied. A three months at this botanic garden would be well spent by any trainee and would not cost a great deal to the host garden other than some time and effort.

Some individual gardens may be prepared to offer shorter specialist courses, perhaps organized once every three years in disciplines for which they have particular expertise. They may include, for example, courses in micropropagation; orchid cultivation; nature reserve and *in situ* plant conservation management; dendrology; greenhouse design, construction and management; seed banking; recording systems and computer methodology. A good example is a training course organized at the Nanjing Botanical Garden Mem. Sun Yat-sen, in co-operation with the BGCS, on garden records systems for Chinese botanic gardens held in November 1990 which was attended by staff from 21 gardens.

One must emphasise that the promotion of training courses for students from the tropics, based in tropical gardens themselves must become a priority for the BGCS in the coming years. Such courses can be hosted by a tropical garden but developed and managed either by the garden itself or with additional assistance, finance and personnel gained through international co-operation with gardens outside of their own region and often from temperate gardens.

The establishment of an international scholarship for staff training is an option that should be considered also. Depending upon its resources, a scholarship could be used to fund one or more candidates from botanic gardens in the developing world each year to spend study periods in an appropriate botanic garden. Of course, such an award would need to be well funded either from an endowment fund for which finance could be sought, organized through commercial of industrial sponsorship or by many smaller but regular subscriptions from supporting organizations.

On site staff training

One should not underestimate the rôle that *on site* garden training can play for staff in their own institution. Visits by botanic garden experts or specialists in one field or other to a tropical botanic garden can help to establish new facilities or programmes and to train staff to manage or administer them. Such visits often help in the creation of important long-term co-operative links between gardens and can be easier to finance than visits in the other direction. As well as that, there are generally as many, if not more, valuable benefits for the consultants themselves as for the host garden and it is rarely difficult to find volunteers to undertake such missions, regardless of whether or not much finance is available.

The rôle of student volunteers

The BGCS has recently started to organize and promote a scheme for student volunteer placements in botanic gardens. This follows our receipt of a significant number of requests or letters from botanical and horticultural students asking if we can recommend botanic gardens for them to work in as volunteers. Some require placements for periods of one to three months. Others are keen to work as volunteers for longer periods. In general, students from temperate regions are keen to work in tropical botanic gardens and vice versa. Some have adequate funding to cover their travel expenses and living costs. Others require an element of financial support. Despite their tender years, many of these students have considerable botanical and horticultural skills and enthusiastic energy that could be of benefit to developing and emerging gardens. We believe that such enthusiasm should be encouraged and, with appropriate screening processes and safeguards, hope to try to help such students to find suitable gardens to visit where their talents can be best used and appreciated. The success of this scheme and its value as a contribution towards tropical garden development will depend upon the degree to which garden directors are prepared to co-operate and inform their students that such an option is available and make volunteer places available.

Training course structures and curricula

In addition to those options I have discussed in this paper, there are many other staff training possibilities worthy of further development as proposed by Simmons et al. (1987), including study visits for garden managers, post graduate courses for scientific staff, internships and summer schools. Although some will be better suited to one institution than to another, all will have a rôle to play in the years ahead.

While discussing the structures that must be put into place for staff training opportunities, we should not forget that close consideration and discussion is urgently required in the formulation of curricula for these training courses. Indeed, a considerable measure of international agreement and co-operation would be valuable in this field. Many training courses in temperate countries are presently not suitable to the needs of participating students from the tropics. Nevertheless they have had to be accepted as the only current option available. Many aspects of the work of a temperate botanic garden are not relevant to those in the tropics. Indeed, most available courses promote technology or horticultural products for their students which will be either extremely expensive or else unavailable in the tropics. As well as that, important subjects such as garden resource and staff management, fund-raising, project and programme development, administration, media manipulation and public education are generally barely touched upon. These should not be regarded as subjects or skills appropriate or only necessary for senior botanic garden managers. In addition, the further subjects of a practical horticultural nature mainly and of particular relevance to tropical botanic gardens, such as tropical pest control, aridity, flooding and irrigation problems and seed recalcitrance, are rarely studied in the the presently available botanic garden staff training courses.

Implementing an international staff training scheme

Following these few years of discussion and careful consideration of the options available for progress since the 1985 Las Palmas Conference, and now that the Botanic Gardens Conservation Strategy is available, the time for action has come. The BGCS is considering plans for the implementation of several ideas put forward for international staff training co-operation. Many gardens have indicated that they will be willing to provide training places, but how many presently do? The BGCS will increasingly be pressing them on their stated willingness to co-operate in this regard and assisting and advising their efforts.

Initially the BGCS hopes to prepare a register of all those gardens prepared to offer training places and courses, expanding upon the information given in the forthcoming International Directory of Botanical Gardens (Heywood et al., 1991) and focussing only on the training options available. It will include information on the facilities available in the garden, whether financial support is possible for trainees and it will provide information on such aspects as accommodation, relevant funding agencies, specific garden interests, etc. It will be maintained by the BGCS as part of its computer database and so can be updated as required. The register will be widely circulated and will act, we hope, as a manual for botanic gardens on how to find suitable training places for their staff, in institutions most suited to their requirements and as a guide to the resources they have available for support of those trainees. The BGCS will act as a central clearing house for these queries and monitor the effectiveness and success of each initiative.

At the same time botanic gardens will be approached directly to outline their plans for the development of their training programmes. A series of pilot training courses in selected gardens will be initiated, both in tropical and temperate countries for which the BGCS will seek funding. Ideally these will be based in a series of regional centres so that travel costs for participants are kept to a minimum. From these pilot projects guidelines and some course outlines will be developed which may be used by other institutions undertaking similar initiatives. An *ad hoc* group of people drawn from BGCS member gardens will advise on these projects and help to ensure that adequate expertise is available for gardens seeking to implement new training programmes. In this regard we will also work closely with other organizations active in this field, especially the International Association of Botanic Gardens (IABG), so that all available resources and expertise can be pooled for the project.

It will be essential for many more enthusiastic individuals to come forward and work with the BGCS to vigorously promote the development of greater international co-operation in staff training in a multitude of gardens throughout the world. In some botanic gardens staff training has been excellently carried out for years but lack of resource sharing, co-ordination and co-operation has hindered the wider application of their expertise.

References

Affolter, J., (1988). California - Costa Rica link-up. *Botanic Gardens Conservation News,* 1(3), 10.

Anonymous, (1988). New Cuban - Spanish agreement. *Botanic Gardens Conservation News,* 1(3), 11.

Heywood, V.H., (1989). *Botanic Gardens Conservation Strategy.* IUCN, WWF and BGCS, Gland.

Heywood, C., Heywood, V.H. and Wyse Jackson, P.S. (1991). *International Directory of Botanical Gardens, V.* Koeltz Scientific Books, Konigstein.

Simmons, J.B.E. and Beyer, R.I., (1987). Cooperation in training and technical liaison between botanic gardens. In: Bramwell, D., Hamann, O., Heywood, V. and Synge, H. (Eds). *Botanic Gardens and the World Conservation Strategy.* Academic Press, London and New York.

Swindells, P., (1987). A Third World horticultural training unit. *Botanic Gardens Conservation News,* 1(1), 20-21.

Resumé

Promouvoir le développement des Jardins Botaniques sous les tropiques dépendra d'un certain nombre de facteurs parmis lesquels les ressources en personnel expérimenté et bien entraîné s'avéront des plus important-s.. La formation de personnel dans les domaines de l'horticulture, la conservation, la gestion et la administration d'un Jardin Botanique, l'education du public et la récolte de fonds aussi bien qu'en botanique pure et en sylviculture sera également d'une importance vitale.

Ce rapport fait le point sur l'état présent de la coopération international dans la domaine de la formation de personnel pour les jardins botaniques sous les tropiques. Il soulignera les réussites et les échecs du passé et indiquera des nouvelles initiatives en cours qui peuventfaire naître des strtégies de progrès pour l'avenir. Bien que des jardins botaniques, surtout dans les régions tempérées, fournissent une précieuse formation aux personnels des jardins botaniques des tropiques, cette formation n'est pas toujours trés axées sur les problemes auxquels ils devont faire face quand ils retourneront dans leurs pays. De plus, il y a très peu de place disponible pour des aspects très importants tels la gestion d'un jardin botanique et collection de plantes.

Nombre des ces problèmes peuvent être résolus ou allégés grâce à une très large coopération internationale, régionale et nationale et par le développement de cours de formation dans beaucoup plus des jardins botaniques qu'ils ne le sont actuellement. De tels cours peuvent être mis sur pied soit par des jardins seuls, soit partagés et organisés par un réseau de jardins botaniques et autres institutions de la région.

Ce rapport insistera sur les éléments de financement nécessaires à de telles avancées et le rôle que le BGCS envisage de jouer dans ce domaine.

Resumen

Promover el desarrollo de los jardines botánicos en los trópicos es una tarea que depende de una serie de factores, entre los cuales el más prioritario es el tener acceso a una plantilla de especialistas. La formación del personal en los campos de la horticultura, conservación, gerencia y administración de los jardines botánicos, educación pública, y finalmente la búsqueda de fondos, son de vital importancia, como lo son el estudio de la botánica pura y la agronomía forestal.

En este artículo se investiga el estado actual de la cooperación internacional en la formación de la plantilla de los jardines botánicos tropicales. Se subrayan los éxitos y fracasos del pasado, y enfoca las nuevas iniciativas que se estan desarrollando actualmente, las cuales quizá sugieran la estrategia para un progreso futuro. Aunque algunos jardines botánicos, en especial los presentes en regiones templadas, presentan el lugar apropiado para la formación del personal de jardines botánicos tropicales, con frecuencia esta formación no se encuentra enfocada adecuadamente hacia los problemas que se les presentarán en el futuro una vez de que retornen a su jardín botánico. Además de esto, existe poca elección de cursos relacionados con un aspecto de tanta importancia como es la gerencia de los jardines botánicos o de las colecciones de plantas.

La mayoría de estos problemas se pueden resolver, o por lo menos aliviar, a través de la cooperación regional, nacional o internacional, y mediante la organización de cursos de formación en más jardines botánicos que los que están organizados en este momento. Los cursos se podrían iniciar bien en jardines botánicos individuales, o bién se podrían compartir y organizar a través de una red de jardines botánicos y otras instituciones de una región determinada.

En el artículo se hace también referencia a algunos aspectos de la financiación requerida para llevar a cabo dicho desarrollo, así como al papel que juega en este aspecto la Secretaría para la Conservación en Jardines Botánicos.

French policy on the protection of wild flora, notably in the overseas territories

J.P. Galland

Le Secrétariat d'Etat à l'Environnement
Neuilly-sur-Seine, France

The French Ministry of the Environment has as one of its primary tasks to look after the conservation of the French natural heritage. Its responsibility extends to all territories governed by France. That is to say the 95 metropolitan departments and 4 overseas departments, which are Guadeloupe, Martinique, French Guiana and Réunion.

French overseas territories have their own local authorities which are responsible for environmental matters, and they will not be included in this paper.

Some figures

One cannot help noticing the very great wealth of plant heritage present on French soil. Metropolitan France is situated at a biographical crossroads. Influences, such as those of the ocean, the Mediterranean and the continental land mass, complicated by the effect of altitude, bring with them a great variety of climates and thus of ecological conditions.

More than 4,500 species of flowering plant occur in metropolitan France, among which there are several hundreds of endemic taxa, although it is difficult to be precise about this number because endemism is a biological notion and not an administrative one. The greater part of the endemics found in the Pyrenees are for example Franco-Spanish endemics. Corsica alone has 130 endemic taxa.

Martinique and Guadeloupe in the Lesser Antilles, are home to about 1,700 species of flowering plant indigenous to Guadeloupe and Martinique, of which 70 are endemics. French Guiana, whose flora is far

TROPICAL BOTANIC GARDENS: THEIR ROLE
IN CONVERVATION AND DEVELOPMENT
ISBN 0–12–346850–7

from having revealed all its secrets, has something like 5,000 species of higher plants. Finally, Réunion has about 800 indigenous species of which 180 are endemic to the island, which represents a very high level of endemism.

The first problem: identification of threatened species

It is a vast undertaking to ensure the conservation of a plant heritage as large as this. The first problem we meet, is of course, the identification of which threatened species require protection measures. Ideally, for each of these we must be able to characterize the degree of rarity, the rate of loss and the amount of threat that is posed to each particular species.

To attain our objective, we need to call on all sorts of skills: those of professional botanists (teachers, university staff, botanic garden staff, etc.), but also, and more and more so, those of amateur botanists whose knowledge is very often extremely valuable.

In order to gather together all these data, to store it, to administer it and to spread information gained as a result of this work, in 1979, the French Minister of the Environment created the Secrétariat de la Faune et de la Flore (SFF), with the assistance of the Muséum National d'Histoire Naturelle. This organization has a network of scientific observers who collect data according to a common national methodology which allows for a coherent efficient approach to data handling, managed with the help of an information system.

SFF gathers several kinds of data for the national heritage inventory from different programmes: the finally completed biogeographical classification of species and their populations; the inventory of natural areas of ecological interest; faunistic and floristic work (the "Z.N.I.E.F.F.") carried out in every region of France; and bibliographical research in order to gather old information to allow for the retracing of trends and historical evolution.

The goal for this work is, of course, to allow continuous monitoring of the evolution of the French natural heritage. To date, the databank of the SFF comprises 4 million observations on species of fauna and flora, the description of 18,000 natural zones and the collection of 28,000 bibliographical references. SFF will also publish the Red Data Book of the threatened flora of France at the request of the Ministry of the Environment.

Legal measures for the protection of species

Having identified what the threatened species are we can then give statuary legal protection by applying the law on the protection of nature (Galland, 1987).

This law enacted on 10 July 1976 defines three methods for the protection for each species:

- a method for complete protection, prohibiting activities which threaten any species (their destruction, ripping out, collection, use, trade),
- a method of partial protection, whereby certain of these activities are made the responsibility of a particular administrative authority,
- a method of protection by local regulations, fixed by the heads of police of each Department, to limit the collection of certain species.

The first of these for complete protection concerns the species which are most threatened. It has been applied to wild species in the following way:

- 400 protected species in the whole of the French metropolitan territories since 1982,
- 61 protected species in Réunion since 1987,
- 42 protected species in Martinique and 36 in Guadeloupe in 1989 (the laws having appeared in the official journal of the French Republic on 3 March 1989).

Apart from this, and to take account of the important distribution of the flora of metropolitan France, the national list of threatened plants is currently being complemented by regional lists, more adapted to biological reality. Four lists have already appeared up to now and numerous others are in preparation.

The method of partial protection concerns species which are rare, but which may be collected under certain conditions. This method has been applied in the case of 34 species, of which some are of interest to the pharmaceutical industry. Under this regime, applications can be made concerning the production, importation, or trading of cultivated specimens of species so protected. There are 400 species completely protected in the whole of metropolitan France. This method allows for trade in such protected species, on condition that they are produced or imported by an authorized organization or establishment.

The method for species protection by local regulations concerns common species which are sometimes gathered on a massive scale. Up to now it has been applied to all species of edible fungi and to fruits of *Vaccinium* spp. It will shortly be extended to a whole list of species frequently gathered in different regions of France.

Legal measures for the protection of wild species

The legal protection of threatened species is only a first step. It is also necessary to adopt measures for the protection of their environment. We have arranged a number of measures in this respect which allow the creation of natural protected areas.

The most powerful of these is, of course, the national park. There are seven national parks in France, the last of these has just been created in Guadeloupe. The creation of national parks protects important natural areas of national importance. These benefit from important funding by the state and from the qualified personnel who care for them and study their natural environment and manage their visitors.

A legal device, more adapted to the conservation of particular natural and remarkable environments is the establishment of nature reserves (réserve naturelle). Nature reserves are established by a law which regulates and forbids those activities which might be to the detriment of the natural development of their fauna and flora, and more generally, which might alter the character of the reserve. Nature reserves themselves protect elements of a natural heritage, which are of national interest. They have a consultative committee and a management body. There are currently 95 of these in existence at the moment. Three of them are situated in the overseas territories, one in Guadeloupe, one in Martinique and the third in Réunion at St. Philippe.

We should also mention the possibility that private owners can seek the designation of "voluntary natural reserve" for their land, either for all or part of their property when such a property is of scientific or ecological interest of a certain level. The designation is granted by the head of the police of the local Department for a period of six years after which time it can then be renewed. The number of voluntary natural reserves is currently 44 in the whole of French territory.

Another legal instrument which is very original and particularly important for the conservation of threatened species is constituted by the "biotope protection laws". This measure is the responsibility of the Prefect of each department and has as its aim the enhancement of the conservation of the environment of a protected species. It is conceived as a flexible measure, easy to put into operation, and can be used for small pieces of land. Currently, there are 130 decrees for the protection of the biotopes in France of which about 50 concern protected plant species.

Another measure for the protection of wild species also merits mention here. It concerns the state forests which are administered Office National des Forêts. When a part of these forests is of particular interest with regard to safeguarding the species it contains, or if it is a biotope of scientific or conservation interest or importance it is possible to designate this as a "biological forest reserve". According to two agreements between the Ministry of Agriculture, the Office National des Forêts and the Ministry of the Environment, the creation of a biological forest reserve is formalized by an administrative document for the forest, which is approved by the Ministry of Agriculture. Biological forestry reserves tend to be completely protected and without human intervention or they are the subject of a particular management regime. The current total of biological forest reserves is 70. In Réunion, 6 of these reserves have been created to date and these represent a total of something in the region of 7,000 hectares.

Finally, to finish this section on *in situ* protection, it is worth mentioning the existence of a public body under the responsibility of the Ministry of the Environment. It is the "Conservatoire de l'Espace Littoral et des Rivages Lacustres" (Conservatory of the shoreline and lakesides). It was created in order to acquire pieces of the coast and land around the great French lakes. These acquisitions are sometimes motivated by the conservation of rare plant species, as was the case in Corsica for the conservation of *Juniperus oxycedrus* subsp. *macrocarpa*.

Ex situ conservation activities

To protect a particular species in its environment is of course the objective which we all have in mind, and this is a top priority. However, we all know that this can often be difficult to achieve. When the numbers of a particular species has reached a critically low level, or when the conservation of its natural biotope appears uncertain, it becomes indispensable to remove it to a safe, preserved, appropriate environment and thus save it from disappearance altogether by cultivating and propagating it.

This *ex situ* conservation phase is particularly critical to manage. A wild population can have a great genetic variability, but this variability has a natural way of limiting itself within a certain time span, if the individuals are in sufficient number and exchange their genes haphazard and without any kind of outside influences. To attempt to conserve a species by cultivating it outside of its own environment is a particularly sensitive undertaking, because the task involved is basically concerned with saving a genetic heritage that is characterized by its variability, and above all, one that is susceptible to being easily altered by human intervention.

It is therefore imperative that we take a certain number of precautions at each phase when undertaking *ex situ* conservation. We have gathered together 6 principal points to be considered when *ex situ* plant conservation is being undertaken:

1. It is important to collect the most representative possible collection of samples of a species, or of the population to be safeguarded

If one only propagates individuals representing a part of the genetic variability of any given population to be conserved, one gets a population which risks being very different from the original one. Geneticists call this the "founder effect".

In addition, if one starts with a small number of individuals, there is a risk that genetic problems may arise in future generations due to the expression of certain deficiencies associated with recessive genes they may contain. Nevertheless, in the case of a very rare species, it is not always easy to get a lot of samples!

2. Prevention of foreign genes entering the genepool of a population being conserved

It is of course well known that when genetically compatible populations are grown close together they inevitably produce spontaneous hybrids between them which would not necessarily have occurred in the wild. The population being conserved therefore "enriches itself" with new genes, which will then modify the nature of the variation in the population. This is to be avoided at all costs if genetic conservation is to be achieved.

3. Prevention of gene losses due to involuntary selection

A population of plants maintained in cultivation risks being put under new selective pressures linked with their new ecological conditions as well as related to the culture practices under which they are grown, which tend, for example, to eliminate individuals which are different from the majority of the rest of the population, and this leads to the "homogenization" of the population. This must be avoided at all costs as it leads to gradual genetic erosion of the population being conserved.

4. Prevention of the loss of genes by genetic derivation

The cultivation of a population of reduced numbers with each generation leads to a loss of a certain number of genes, due to what can be described as a form of "hazardous lottery" of the genes which are transmitted to the following generation. This phenomenon is called "genetic derivation" and leads to a dangerous homogenization of the population.

It is therefore imperative to cultivate populations in sufficient numbers. The minimum threshold is difficult to be precise about, generally speaking, because it depends on the reproductive strategy of the plant in question.

5. Choice of the best adapted mode of cultivation

We have seen that, generally speaking, one species of plant is composed of several populations. When collecting samples of the same species, but from different origins, it is not always easy to know precisely when one is dealing with different populations. One can therefore be tempted to cultivate and to conserve separately these populations with different origins. Thus one constitutes a collection of individual origins.

But this is not easy and difficult to put into practice. It is wiser sometimes to regroup the different origins of the same species in a new population. This mode of cultivation is the simplest to put into practice and is less sensitive to the genetic erosion than individual lines, but it much more sensitive to the selection process. It is, therefore, not always easy to choose the most appropriate mode of cultivation for the conservation task in hand.

6. The maintenance of the same vigilance in reintroduction programmes

In cases of threatened wild species, we make a distinction between 3 types of replanting of plants into their natural environment:

- the strengthening of a population in a location where the taxon under consideration is still present, but where it has become rare.
- the reintroduction in a location where the taxon has existed in the past, but from where it has disappeared.
- the introduction to a location which is ecologically favourable, but where the taxon under consideration does not seem to have existed in the past.

In each of these three cases it is imperative to observe strict rules governed by genetic considerations concerning the populations which we have just referred to (Bournérias & Olivier, 1989). For example, in the case of the strengthening of a population by the addition of further individuals, it is vital only to use plant material derived from the locality where the reintroduction programme is being undertaken.

In the case of reintroduction, or simply introduction, we must use material which comes from a location which is geographically close so that the risk of altering genetically compatible populations near the introduction site is reduced. It is apparent that all of these precautions can only be carried out if a permanent effort is made to monitor the introduction and if adequate staff training is given in these methods.

In botanic gardens, there is often a great number of species represented by just a few individuals and of which some are interfertile. In some cases we are talking about conditions which are very far removed from those necessary for genetic conservation. This is why the French Ministry of the Environment saw fit to create a category of institution for those establishments wishing to be involved in the genetic conservation of threatened wild plants and to make conservation one of their priorities.

The designation of National Botanical Conservatory (Conservatoire Botanique National) was defined by a decree issued on 12 April 1988 (and which appeared in the official journal of the French Republic of the 14 April 1988). This designation is given by the Ministry of the Environment in return for the observance of a "cahier des charges" (book of responsibilities) which outlines constraints which, most notably, promote the measures listed previously undertaken to ensure the genetic conservation of endangered species (Projet de cahier des charges pour les conservatoires botaniques nationaux. Document disponible à la Direction de la Protection de la Nature du ministère de l'environnement, 14 Bld. Général Leclerc, F 92524 - Neuilly/Seine Cedex.).

The national botanical conservatories have three aims:

1. The identification of rare or threatened wild flora of a given region or a particular group of species and their distribution.
2. The conservation by all appropriate means (that is to say, both *in situ* protection and *ex situ* conservation) of threatened taxa.
3. Education of the public and administrators of the relevant region of the country by the conservation of the plant heritage.

Any establishment (for example, a botanic garden) which wishes to obtain this designation, must choose a specialization, that is to say, an area of activity which is limited to one part of French territory or to a group of particular species (for example, endemic species, medicinal plants etc.). Its application will be studied by a scientific committee which has as one of its most important rôles the job of producing the "book of responsibilities" which the conservatory in question must respect in carrying out its particular specialization.

Currently, the Ministry of the Environment is examining three designation applications which are from the Conservatoire Botanique de Brest, the Conservatoire Botanique de Porquerolles and the Conservatoire et Jardins Botaniques de Nancy. These are precisely the same institutions that the Ministry of the Environment established a few years ago, and whose rôles will thus be strengthened and formalized. They will soon be joined by other new institutions: certain French botanic gardens are already showing interest in seeking such designation.

The aim of the Ministry of the Environment is to set up a network of co-ordinated establishments working according to a common methodology, and constituting real partners for those authorities which are responsible for the protection of nature in the national territory.

Conclusion

France has chosen to link its threatened plants conservation programme with the conservation of the national heritage. A priority has been given to achieving the preservation of the genetic variability of species. While this may be difficult to achieve, we do not see that we have any alternative.

Undoubtedly, to consider a species as an ensemble of genes which it is convenient to conserve, is not a very poetic notion, but we must not forget that the wild flora constitutes not only a heritage which it is incumbent upon us to conserve for ethical reasons, but also it constitutes a resource which will allow humanity to respond to real needs in many ways in the future.

Humanity needs to rely on a natural environment that is not only rich and varied in the number of species it contains, but also to be able to use the variability found within each species.

References

Beaufort, F.de et Maurin H., (1988). *Le secrétariat de la faune et de la flore et l'inventaire du patrimoine naturel*, 2 vol., Paris. (disponible au SFF, Museum National d'Histoire Naturelle, 57, rue Cuvier, F 75231 Paris Cedex 05.

Bournérias, M. & Olivier, L., (1989). Déontologie et méthololologie applicables aux renforcements, réintroductions et introductions de flore dans le milieu naturel. In: *Actes du colloque: Plantes sauvages menacées de France, bilan et protection,* Brest, 8-10 Octobre 1987, 379-385. Bureau des Ressources Génétiques, Paris.

Galland, J.P., (1987). Les instruments juridiques de protection de la flore sauvage en France. In: *Actes du colloque: Plantes sauvages menacées de France, bilan et protection,* Brest, 8-10 octobre 1987, 233-242. Bureau des Ressources Génétiques, Paris.

Resumé

La France, avec notamment ses Départements d'Outre-mer, possède un patrimoine végétal très riche comportant de nombreuses espèces endémiques.

Le ministère français de l'environnement a créé avec le Muséum National d'Histoire Naturelle unorganisme chargé de collecter et de gérer l'ensemble des données sur le patrimoine naturel (espèces et milieux): le secrétariat de la faune et de la flore.

Lorsque les espèces menacées sont identifiées, il est possible de leur donner un statut de protection juridique. La loi française sur la protection de la nature permet trois régimes de protection selon le degré de menaces: protection intégrale, protection partielle et réglementation locale de la cueillette.

Par ailleurs, pour protéger le milieu de vie des espèces menacées, nous disposons d'un arsenal juridique permettnt la création d'espaces protégés (parc national, réserve naturelle, réserve naturelle volontaire, arrêté de protection de biotope, réserve biologique forestière) ou l'Etat de terrains sur le littoral.

Mais la protection in situ est très souvent insuffisante voire impossible. Il est alors nécessaire de faire appel à la conservation ex situ. Il s'agit d'une phase très délicate car dans la nature les plantes échangent leurs gènes entre elles au sein d'une même population, au cours de la reproduction sexuée.

Pour conserver dans de bonn es conditions une espéce hors de son milieu, les lois de la génétiue des populations nous imosent des règles strictes:

- collecter un échantillonnage représentatif de la variabilité de l'espèce,
- éviter l'arrivée de nouveaux gènes par hybridation involontaire,
- éviter la perte de gènes par sélection ou par dérive génétique,
- conserver séparément les différentes populations d'une même espèce lorsque cela est possible et souhaitable.

En France, ces tâches délicates sont conduites par des établissements spécialisés: les "conservatoires botaniques nationaux". Ces établissements sont agréés par le ministère de l'environnement en échange du respect d'un cahier des charges trè strict.

l'objectif du ministrè de l'environnement est de constituer et de coordonner un réseau de conservatoires botaniques assurant dans les meilleurs conditions la conservation du patrimoine génétique végétal sauvage pour le bénéfice des générations futures.

The work of WWF-France with regard to botanic gardens in the West Indies

C. Sastre

WWF-France

Summary

WWF-France supports many actions in Metropolitan France and more recently in overseas Departments (D.O.M.). Since 1986, WWF-France has collaborated with the "Association Saintoise de Protection du Patrimoine" (ASPP) and "Conservatoire des Jardins et Paysages" to make educational displays for the understanding of Les Saintes natural environment and the Jardin Exotique du Fort Napoléon (Terre de Haut, Archipel des Saintes, Guadeloupe). That garden specializes in succulent plants and illustrates to the public various botanical families. It is now becoming a conservatory of local endangered species, especially Orchidaceae and Cactaceae and including reproduction *ex situ* to provide plants to reintroduce into the wild. In Martinique, contacts exist with the Jardin de Balata, Fort de France. Orchid Societies also help in maintaining native biological diversity.

Introduction

Since its creation in 1973, WWF-France has supported many projects in metropolitan France (the Camargue reserves, the purchase of marshes in Lorraine, the protection of seals in Britain, the reintroduction of lynx into the Vosges, etc.) and overseas: tortoises in Guyana and support for the Jardin Exotique du Fort Napoléon (in Guadeloupe). In the WWF-France magazine "Panda" all members are informed about different actions undertaken, and it contains articles on subjects some of which

TROPICAL BOTANIC GARDENS: THEIR ROLE IN CONVERVATION AND DEVELOPMENT
ISBN 0–12–346850–7

are of interest to us here. Number 35 for example, concentrates on Guyana and Number 39 is devoted to the tropical forest.

These WWF-France activities are of two types:

- the collection of funds, without which nothing can be achieved, for financing its contribution to WWF-International and for programmes planned by the administrative council on the basis of advice from the scientific council,
- the bringing about of activities which advance the World Conservation Strategy, with the following aims:

a) to study the threats which endanger environments and species;

b) to alert governments and decision makers to these threats and to raise the consciousness of public opinion;

c) to work in collaboration with organizations concerned with these issues, and with local associations for the protection of nature.

Let us now look at some of our botanical activities in Guadeloupe and Martinique, and particularly, in Les Saintes Islands.

Jardin Exotique du Fort Napoléon

The 4 June 1984 saw the inauguration of this garden after two years of work by the "Association Saintoise de Protection du Patrimoine" (ASPP), supported by the Commune de Terre de Haut. The building of this Fort was finished under Napoléon III, was never used militarily and was then abandoned at the beginning of the 20th Century. It was then part restored and the building became a museum, and the surrounding grounds became a botanic garden, where succulents largely featured, a reason for its twinning with the Jardin Exotique de Monaco.

Educational posters on the subject of discovering the wild in the island were requested from the "Conservatoire des Jardins et Payages". Moral and financial support was requested from WWF-France, and they accepted to participate in the project. I was sent in August 1985 to start the poster project. At first with little knowledge of the local flora, botanical (and zoological) research had to be undertaken, carrying out a floristic inventory and, with the assistance of numerous local people, recording the local names and uses to which plants were put.

I very quickly realised that the Terre de Haut had a very characteristic fauna and flora specific to it. I discovered the presence of two species of iguana, the existence of a dry relict forest with *Mastichodendron foetidissimum* (Jacq.) Cronq. (Sapotaceae) and *Hymenea courbaril* L. (Caesalpinaceae), an epiphytic orchid site near the summit of Chameau at 309 m. that is rare in Guadeloupe, and the existence of the famous succulent, “tête anglaise”, *Melocactus intortus* (Mill.) Urb. (Cactaceae), endemic to the dry Lesser Antilles.

In discussions with the local population and through the consultation of local archives it became apparent that there was a very rich history, especially military. The construction of many buildings (batteries, towers, forts) had a great impact on the natural environment. One needs to have a knowledge of these aspects of history to gain an insight into the reasons for the present day vegetation patterns in Les Saintes.

The population of approximately 1,500 inhabitants of Terre de Haut name many plants differently from the way they are named in Guadeloupe, where there are 250,000 inhabitants and where a significant ethnobotanical influence can be noted in local names.

In the light of the fact that many botanical discoveries had never been registered in Les Saintes (and were even unknown to many local people), for example, the coastal plum (*Ximenia americana* L.), an exhaustive inventory of the flora of Terre de Haut had to be undertaken. More than 300 species, including those in gardens were listed.

The orchids and the “tête anglaise” cacti had become very rare in recent years in Guadeloupe, and its dependencies, because of numerous gathering missions for plants destined for Europe and North America. In addition, the Jardin Exotique du Fort Napoléon started to play the rôle of botanic conservatory in a sense of what is contained in the law called decree 88-352 of the French Republic. The nursery now features “tête anglaise” cacti found and abandoned throughout the little paths of the island. Seeds were sown with success, and an attempt made to replant the species in natural wild sites.

Contacts were made with the “Association Guadeloupéenne d’Orchidophilie” (AGO) of which one of the members worked at the CIRAD and has access to an *in vitro* propagation centre, with the aim of producing native orchids in Les Saintes. Already, plants gathered, either on the ground or collected by visitors are cultivated on several trees in the garden.

Little nature trails have been opened, avoiding very sensitive areas or highlighting species which are severely threatened and which therefore, are not easily visible, except in the garden. One site, at Morne Morel, is being prepared to receive plant reintroductions to the wild from the garden.

The posters, whose text has been completed and is at the checking stage, is in the hands of members of the ASPP. There are ten posters on the topics as follows: general, the coast, the forest, *Cactus* and succulents of the cliffs, fruits and vegetables, garden flora, reptiles and amphibians, birds, botanical trails and the Fort Napoléon.

The illustrations of the posters enlarged and framed will serve as a basis for an ecomuseum which will be located in the central building of the Fort. This will come into being as an educational tool which will also play the rôle in protecting species and spaces through local initiatives.

We hope that with this structure the garden will be able to join others in a network of botanic gardens of the Caribbean. We hope too that wild sites of great floristic and faunistic interest such as Chameau and the Grand Ilet will be the subject of further habitat protection laws, thus ensuring that they will have legal protection.

In 1990 we have before us a project to organize a botanical seminar in Ile de Saintes on the dry vegetation of the Antilles and its protection. For this, we hope to define the work before us and to present it to the international scientific community. On this occasion the scientific council of the garden will be established.

Other activities

In Martinique contacts have been made with the Jardin de Balata, situated in the hills of Fort the France. Its owner, Jean-Philippe Thoze, has made a wonderfully pleasing garden there, mixing exotic and indigenous plants of which some have been saved from the bulldozers of the Office National des Forêts, or those of the Department de Ponts et Chaussées. There had been numerous epiphytic members of the Orchidaceae and Bromeliaceae belonging to the genera *Oncidium* and *Aechmea*, which unfortunately grew in an area destined to become a forest of mahogany, *Swietenia macrophylla* King.

Marie-Claude, the wife of M. Thoze, has decorated their house with botanic drawings and various fruits and floral bouquets of her own composition, was thus giving homage to the plant world also.

But for WWF-France to be able to invest in this garden, there would have to be a scientific council, and above all, strict appreciation that exotic species introductions, for example, bamboos, can become veritable plant pests.

Orchid enthusiasts from Guadeloupe and Martinique also play a rôle in all this, a rôle which has significance for the preservation of biological diversity. This is one of WWF's highest priorities, and we are happy to collaborate in their exhibitions and provide panels and small items which WWF can sell.

Conclusions

Our experience with West Indian gardens has led us to certain conclusions.

Botanic gardens and conservatories can take their place alongside other methods with which we can undertake the study and investigation of the biology of plants. These same structures can equally contribute to protection and the survival of those species which are the most threatened.

Botanic gardens and conservatories can protect species, but the conservation of their environment is also a priority which we must not lose sight of amongst those who are working in botanic gardens.

The choice of species to be protected will have to be made in many cases. The publication of red data books often masks the lack of knowledge about a considerable number of tropical species. In addition, it would seem indispensable, that botanic gardens, and above all conservatories, acquire scientific councils or committees which will, not only promote the conservation of priority species, but will also ensure that follow-up operations are undertaken.

The final goal of protection is to conserve ecosystems and so the rôle of conservatories in preserving species must be considered as an intermediate stage, helping to preserve them during difficult times.

Resumé

WWF-France appuie diverses actions en France métropolitaine et plus récemment dans les Départements d'Outre-Mer (D.O.M.). Depuis 1986, WWF-France collabore avec l'Association Saintoise de Protection du Patrimoine et le Conservatoire des Jardins et Paysages à la réalisation de fiches didactiques portant sur la découverte des milieux naturels saintois et du Jardin Exotique du Fort Napoléon (Terre de Haut, Archipel des Saintes, Guadeloupe). Ce jardin, spécialisé sur les plantes succulentes montre au public diverses familles botaniques du monde. Maintenant il devient un Conservatoire d'espèces locales en danger: Orchidacées et Cactacées, incluant leur reproduction *ex situ* afin de mettre les jeunes plantes *in situ*. En Martinique des contacts ont été pris avec le Jardin de Balata, Fort de France. Les Sociétés d'orchidophilie concourent au maintien de la diversité biologique.

Resumo

WWF-France apoya muchas acciones en Francia metropolitana y mas recientemente en los Departementos de Ultra-Mar (D.O.M.). Desde 1986 WWF-France colabora con la "Association Saintoise de Protection du Patrimoine" y "Conservatoire des Jardins et Paysages" para realizar fichas didácticas para la descubierta de los medios naturales de los Santos y del Jardín Exótico del Fort Napoléon (Terre de Haut, Archipel des Saintes, Guadeloupe). Este jardín especialiado con plantas succulentas presenta al público varias familias botánicas del mundo. Ahora llega al nivel de un Conservatorio de especies locales en peligro: Orquidáceas y Cactáceas, incluyendo reproduccion *ex situ* para poner jovenes plantas *in situ*. En Martinica, contactos son tomados con Jardín de Balata, Fort de France. Las Sociedades de Orchidophilistas ayudan para mantener la diversidad biológica.

Part 2

Tropical Botanic Gardens: Case Studies

The ethnobotanical garden and tropical plant conservation

Brien A. Meilleur

Amy B.H. Greenwell Ethnobotanical Garden
Bernice Pauahi, Bishop Museum, Captain Cook, USA

Summary

While not focused specifically on the maintenance of plant biodiversity, ethnobotanical gardens can nevertheless make significant contributions to endangered species conservation in the tropics. By documenting, then cultivating plants of ethnobotanical interest to indigenous peoples, and then by educating the public about traditional plant use, ethnobotanical gardens combine the preservation and the interpretation of cultural and natural history into single events. It is argued in this paper that this is an especially useful approach in the tropics where many one-time economic species are now seriously threatened, and where greater support for plant conservation from within the tropical countries themselves is needed.

Introduction

Less directly conservationist as living plant repositories than most modern botanic gardens (e.g. Heywood, 1987: 16-17; Smith, 1985), ethnobotanical gardens have the potential to be effective in tropical plant conservation in other ways. The ethnobotanical garden's focus on the utility and significance of plants to people, rather than on the taxonomic study of plants or on germplasm protection per se, can and should be used to advantage in the effort to maintain biodiversity in the tropics. Rarely are significant numbers of plant species being lost in areas not traditionally inhabited by indigenous or peasant peoples and the ramifi-

TROPICAL BOTANIC GARDENS: THEIR ROLE
IN CONVERVATION AND DEVELOPMENT
ISBN 0—12—346850—7

cations of this fact are increasingly becoming fundamental considerations in conservation and development projects in tropical regions. Through well conceived programmes in research, education, conservation and community development the ethnobotanical garden could play several key rôles in this evolution, any of which might substantially contribute to maintenance of tropical plant biodiversity. Greater emphasis should be placed in tropical regions on creating local and regional ethnobotanical gardens with interdisciplinary staffs of anthropologists, botanists and ecologists that specialize in traditional plant use and resource management, while wide-ranging exotic plant collections with little or no relevance to local or regional ecological and economic conditions should be deemphasized.

Research

Like the botanic garden, research must be a fundamental part of the ethnobotanical garden (for early ideas on the research value of ethnobotanical gardens see Harshberger, 1896: 153-154). However, the ethnobotanical garden shifts the focus from such things as acclimatization, western-oriented economic botany, and taxonomy, to the documentation and interpretation of the cultural significance of plants and to the analysis of their rôle in human adaptation to diverse environments. Indigenous peoples have generally been very effective stewards of the biodiversity around them, and sustained use of plant and animal products is always a key feature in their economic and ecological success (e.g. Williams & Hunn, 1982; the "Declaration of Belem" by the International Society of Ethnobiology, 1988). Within the last 20 years there has been growing recognition of the sophistication and, in many cases, the empirical soundness of traditional human ecological and biological knowledge (e.g. Alcorn, 1984; Freeman & Carbyn, 1988; Glick, 1987; Meilleur, 1986; Nietschmann, 1985).

Scientists, planners and government officials are increasingly turning to ways in which this knowledge can be applied to development and conservation in more remote parts of the world, and on ways in which indigenous peoples can become involved to a greater degree in both the planning and the outcome of such projects (e.g. Cohn, 1988; Posey, 1983; Posey et al., 1984; Rhoades & Booth, 1983). Locally/regionally oriented ethnobotanical gardens in the tropics can become players in this important process by motivating and coordinating research into traditional plant and land use, and by widely disseminating results. In this way valuable native management principles and traditional plant uses have greater chances of being considered in decision-making processes affecting natural resources, decisions which today are often entirely removed from the hands of indigenous and peasant peoples.

In afforestation projects, for example, the replacement of exotic trees with little or no economic value (e.g. the afforestation pattern in Hawai'i during the first half of this century (LeBarron, 1970) by woody species with indigenous cultural importance, would not only be economically valuable to native peoples and politically well received by them, it could in some cases be the precursor to the reestablishment of natural forests. On a smaller scale, such forests might be intensively managed for sustained yields, thus reducing gathering pressure by indigenous people on natural forests (e.g. Alcorn, 1984: 393). Local/regional ethnobotanical gardens would be well suited to participate in applied research programmes of this kind.

More punctually, research generated by ethnobotanical gardens can be useful in establishing priorities in the conservation of endangered or threatened tropical species (e.g. Hamann, 1987: 39). In the absence of scientific knowledge of a threatened plant's economic value, one of the best ways to estimate its potential for wider medicinal or industrial use - and thus the urgency for its protection - is through a thorough evaluation of its use by indigenous peoples (e.g., Cox et al., 1989; Spjut, 1985). Local/regional ethnobotanical gardens could also conceivably become centres where the study of the nutritional value and/or the physiological activity of such plants could be coordinated, with the results restituted quickly to the original indigenous discoverers and to local governments, rather than only to western science and industry, that still often the case today.

Education

Education is as critical to the ethnobotanical garden in tropical plant conservation as it is to the botanic garden (e.g. Eloff, 1987: 87-88). However, by creating interpretative programs which simultaneously explain local/regional natural and cultural history, ethnobotanical gardens may potentially reach a wider audience than do botanic gardens. A goal of ethnobotanical gardens in the tropics might be to act as catalysts, facilitating exchange of information and perspectives relating to conservation and development among indigenous and peasant peoples, urban dwellers, scientists, planners and government officials. Each group has much to learn from the others. Indigenous peoples could be apprised in more humane ways to the larger economic and developmental objectives of nation-states while being exposed to scientific ecological principles and to the ofttimes abstract linkage between local resource use and wider conservation problems which may be exacerbated by such activities (see, e.g., Tuan, 1968; for a dramatic prehistoric example of indigenous environmental degradation in Polynesia, see McCoy, 1979: 159-161).

On the other hand, scientists, planners and government officials would not only benefit from learning about the importance of plant resources to indigenous peoples, and of the emotions and even tragedy sometimes associated with their loss, but in practical terms they may find their own development or conservation efforts enhanced through the incorporation of indigenous resource management principles (e.g. Freemann & Carbyn, 1988: "Preface"; Lamb, 1987: 447+; Ruddle & Johannes, 1985: 1-4). As Alcorn has put it (1984: 391), "policy makers need to inform more fully their "top-down" policies with knowledge of the "bottom" and "external" sectors." Through public programs which highlight the biological uniqueness and the local cultural-historical utility of native plants, urban dwellers - often still linked culturally or in spirit to rural ethics stressing plant and animal importance - can be stimulated to demonstrate greater national or ethnic pride in their biotic and cultural heritage, while they are sensitized to the value of maintaining genetic diversity (e.g. Boden & Boden, 1987: 72; Cohn, 1988: 450, 452). Such awareness could act to nurture stronger conservationist public opinion and increase pressure on governments to move in directions generally beneficial to both *in situ* and *ex situ* plant conservation efforts.

Conservation

Ethnobotanical gardens are more directly focused on the preservation of the cultural traditions associated with plant use and management than with the conservation of the plants in and for themselves. Nonetheless, ethnobotanical gardens are composed of living collections, and to the extent that threatened or endangered species are or were significant to indigenous peoples, there will be overlap between the goals of *ex situ* plant conservation and ethnobotanical interpretation. At Bishop Museum's Amy B.H. Greenwell Ethnobotanical Garden, in Hawai'i, many of our collections - especially those species once common in lowland dry forest ecosystems - are today also exceedingly rare species. By bringing them into the garden and simultaneously educating the public about their cultural importance and their biogeographical rarity, we feel that their *in situ* conservation status can be significantly enhanced as well.

Community development

It is now widely recognized that tropical plant conservation is inextricably associated with local community development (see, e.g., Cohen et al., 1990; Hamann, 1987). The goal, stated in its most simple terms, is to reduce inefficient and wasteful extensive use of tropical plants and ecosystems and to accentuate intensive, sustainable use of these same or other well-adapted resources on smaller scales. This increasingly accepted objective, when combined with the goal of creating new economic opportunities for indigenous peoples and finding solutions to their growing energy needs, is seen by many as the only realistic option which can reduce the current pressure on wild plants and ecosystems by growing numbers of urban and rural peoples alike (e.g. Synge, 1987: xxiv). Through what has come to be called "social forestry" and other programmes, some nation-states in the tropics are increasingly turning to smaller scale development based on what are perceived to be the best of local/regional traditional resources, technologies, and management strategies (e.g. Fox, 1989), rather than on the importation of exotic cash crops and costly technologies often adapted to temperate and other ecologically inappropriate settings.

With an interdisciplinary staff of anthropologists, botanists and ecologists engaged in active field and experimental research programmes, the ethnobotanical garden would be in a unique position to advise state agencies on the ecological soundness of traditional resources, technologies, and management in local development. It would also be well positioned to modify these or other technologies and germplasms, then disseminate them to indigenous peoples in accordance with local needs, whether these be for subsistence or for commerce (e.g. Domingo, 1987; Harris & Kapoor, 1988). In this way, the ethnobotanical garden could contribute to the realization of the third option in local community development proposed in the World Conservation Strategy-2 draft: indigenous and peasant peoples entering the 20th and 21st centuries combining appropriate modern technology with traditional cultural values, rather than being constrained to "return to ancient ways" or to abandon "subsistence altogether [to] become assimilated into the dominant society" and economic system (IUCN, 1988). It is increasingly imperative, as Davis (1977: xviii), and others have stressed, for sound indigenous management principles to be "...translated into strategies of public action, account ability and change" (cited in Lamb, ibid.: 450). A review of the literature shows that several botanic gardens, especially in Asia, have already incorporated such ethnoecological rôles into their operational plans (Dejun, 1983: 226; Zaifu, 1987: 249 +).

Conclusion

I have argued in this paper that ethnobotanical gardens can make significant contributions to tropical plant conservation through well conceived research, education, conservation and community development programs. Many of the ideas and initiatives presented here are not new, though the proposal to embody many of them concretely in the actions of local/regional ethnobotanical gardens is. If we must accept Raven's (1987) recent dire predictions about tropical biodiversity loss in the next several decades, we must also continue to explore new possibilities which may help us to avoid this scenario. The growing consensus that nation-state administrators, scientists, and non-government organizations must increase their co-operation with indigenous or peasant peoples toward this end suggests that the ethnobotanical garden may be a very effective vehicle for implementing such a synergistic approach. The emphasis here on creating or shifting means to ethnobotanical gardens should not be construed as competitive with either *in situ* conservation programs or *ex situ* botanic garden conservation. Rather, it should be seen as a complementary effort within the World Conservation Strategy aimed at stimulating and facilitating co-operation in plant conservation among all segments of concerned society, and especially within the tropical countries themselves.

References

Alcorn, J. (1984). Development policy, forest, and peasant farms: reflections on Huastec-managed forests' contributions to commercial production and resource conservation. *Econ. Bot., 38*(4), 389-406.

Boden, R.W. & Boden, E.A. (1987) Botanic gardens and community education in Australia. In: *Botanic Gardens and the World Conservation Strategy* eds. Bramwell, D., Hamann, O., Heywood, V. Synge, H., pp. 67-74. Academic Press, London & New York.

Cohen, J. & Potter, C. (1990). Genetic resources and conservation and utilization in the context of international development. *Diversity 6*(1), 18-21.

Cohn, J. (1988). Culture and conservation. *Bioscience 38*(7), 450-453.

Cox, P.A. Sperry, L.R., Tuominen, M. & Bohlin, L. (1989). Pharmacological activity of the Samoan ethnopharmacopoeia. *Eco. Bot. 43*(4), 487-497.

Davis, S. (1977). *Victims of the Miracle: Development and the the Indians of Brazil.* Cambridge University Press, Cambridge.

Dejun, Y. ed. (1983). *The botanic gardens of China*. Science Press, Beijing.

Domingo, T. (1987). Ethnobotany of the Buhi Agta of the Philippines. *Tradition, Conservation & Development*, 5, Occasional Newsletter of the Commission on Ecology's Working group on Traditional Ecological Knowledge, IUCN, Gland.

Eloff, J. (1987). Botanic gardens and education in South Africa. In: *Botanic Gardens and the World Conservation Strategy*, eds. Bramwell, D. et al., pp. 85-100. Academic Press, London & New York.

Fox, J. (1990). Social forestry planning: searching for a middle way. *East-West Center, Environment and Policy Institute Working Paper*, 22, Honolulu.

Freeman, M. & Carbyn L. eds. (1988). *Traditional Knowledge and Renewable Resource Management in Northern Regions*. IUCN Commision on Ecology & the Boreal Institute for Northern Studies, Edmonton, Alberta.

Glick, D. (1987). Natural resource management by native peoples in the Americas. *Tradition, Conservation, and Development* 5, Occasional Newsletter of the Commission on Ecology's Working Group on Traditional Ecological Knowledge, IUCN, Gland.

Hamann, O. (1987). The IUCN/WWF Plants Conservation Programme in action. In: *Botanic Gardens and the World Conservation Strategy*, eds. Bramwell, D. et al., pp. 31-43. Academic Press, London & New York.

Harris, W. & Kapoor, P. (1988). Summary report of international workshop on ethnobotany, Nga Mahi Maori O te hao Nui a Tane. *Commonwealth Science Council Technical Report* (New Zealand).

Harshberger, J. (1896). The purposes of ethno-botany. *Bot. Gazette, 21*, 146-154.

Heywood, V.H. (1987). The changing rôle of the botanic garden. In: *Botanic Gardens and the World Conservation Strategy*, eds. Bramwell, D. et al., pp. 3-18. Academic Press, London & New York.

International Society of Ethnobiology (1988). Declaration of Belem. *Jour. of Ethnobiology, 8*(1), v.

IUCN (1988). World Conservation Strategy-2. *Tradition, Conservation, and Development* 6, Occasional Newsletter of the Commission on Ecology's Working Group on Traditional Ecological Knowledge, IUCN, Gland.

Lamb, F.B. (1987). The rôle of anthropology in tropical forest ecosystem resource management and development. *The Journal of Developing Areas 21*, 429-458.

LeBarron, R. (1970). The history of forestry in Hawaii - from the beginning through World War II. *Aloha Aina*, April, 1970.

McCoy, P. (1979). Eastern Island. In: *Prehistory of Polynesia*, ed. Jennings, J., pp. 135-166. Harvard University Press, Cambridge, Mass.

Meilleur, B. (1986). *Alluetain Ethnoecology and Traditonal Economy: The Procurement and Production of Plant Resources in the Northern French Alps*. Ph.D. dissertation (anthropology), Univ. of Washington, Seattle, Wa.

Nietschmann, B. (1985). Torres Strait islander sea resource management and sea rights. In: *Traditional Knowledge and Management of Coastal Systems in Asia and the Pacific*, ed. Ruddle, K. & Johannes, R., pp.125-154. UNESCO.

Posey, D. (1983). Indigenous ecological knowledge and development. In: *The Dilemma of Amazonian Development*, ed. E. Moran, pp. 225-257. Westview Press, Boulder, Co.

Posey, D., Frechione, J., Eddins, J. & Francelino da Silva, L. (1984). Ethnoecology as applied anthropology in Amazonian development. *Human Organization, 43*(2), 95-107.

Raven, P. (1987). The scope of the plant conservation problem world-wide. In: *Botanic Gardens and the World Conservation Strategy*, eds. Bramwell, D. et al., pp. 19-30. Academic Press, London & New York.

Rhoades, R. & Booth, R. (1983). Interdisciplinary teams in agricultural research and development. *Culture and Agriculture, 20*, 1-7.

Ruddle, K. & Johannes, R. eds. (1985). *The Traditional Knowledge and Management of Coastal Systems in Asia and the Pacific*. UNESCO.

Smith, N. (1985). Botanic gardens and germplasm conservation. *Univ. of Hawaii Harold L. Lyon Arboretum Lecture*, 14, Univ. of Hawaii Press, Honolulu.

Spjut, R. (1985). Limitations of a random screen: search for new anticancer drugs in higher plants. *Econ. Bot. 39*(3), 266-288.

Synge, H. (1987). Introduction. In: *Botanic Gardens and the World Conservation Strategy*, eds. Bramwell, D. et al., pp. xxi-xxxv. Academic Press, London & New York.

Tuan, Y. (1968). Discrepancies between environmental attitude and behaviour: examples from Europe and China. *Canadian Geographer, 12*(3), 176-191.

Williams, N. & Hunn E. eds. (1982). *Resource Managers: North American and Australian Hunter-Gatherers.* Westview Press, Boulder, Co.

Zaifu, X. (1987). The work of Xishuangbanna Tropical Botanical Garden in conserving the threatened plants of the Yunnan tropics. In: *Botanic Gardens and the World Conservation Strategy*, eds. Bramwell, D. et al., pp. 239-254. Academic Press, London & New York.

Resumé

Bien que le jardin ethnobotanique ne soit pas spécifiquement fondé sur le maintien d'une diversité biologique, il peut néanmoins contribuer à la conservation des plantes menacées d'extinction sous les tropiques. A travers ses activités de documentation, cultivation, et education relatives à l'utilisation traditionelle des plantes, le jardin ethnobotanique combine l'interprétation et la préservation de l'histoire naturelle et culturelle. On propose que cette approche sera particulièrement utile dans les efforts de protection de la nature dans les pays tropicaux où de nombreuses plantes d'importance économique pour les populations indigènes sont aujourd'hui sérieusement menacées et où un mouvement accru vers des actions de conservation est désiré.

Resumen

Los jardines etnobotánicos pueden contribuir de forma significativa a la conservación de especies en peligro de extinción en los trópicos, aunque no estén específicamente centrados en la conservación de la diversidad vegetal. Combinan en un solo aspecto la preservación, interpretación cultural e historia natural, informando de ello a la gente local, cultivando las plantas de interés etnobotánico, e informando después al público sobre su uso tradicional. En el artículo se discute la utilidad de este acercamiento en los trópicos, donde numerosas especies que en su tiempo tuvieron importancia económica actualmente se encuentran seriamente amenazadas, así como la necesidad de un mayor apoyo para la conservación de las plantas por parte de los propios países tropicales.

The Botanic Garden of Quindío: five thermic zones of Andean flora

Alberto Gómez Mejía

Jardín Botánico del Quindío
Armenia, Colombia

Summary

The Province of Quindío is less than 2,000 sq. km. (0.17% of the total of the nation) and is situated in the centre of Colombia. Although colonization took place only 150 years ago, the area is economically very productive with intensive coffee plantations. It is also an area of high biological diversity. It is estimated that a total of 4,000 species of flowering plants are present in the region, although the number of species properly recorded is not greater than 1,000.

The Botanic Garden of Quindío is a non-governmental and non-profit making institution. Situated in the town of Filandia at 1,950 m., it contains 10 hectares of reasonably well preserved natural forest. The Garden has also been able to raise US$155,000 to acquire a further 14 hectares of natural forest, located close to the city of Calarcá, at 1,490 m., in an area of high population and intensive coffee growing where wild areas have been practically devastated. It is intended to incorporate similar areas, at several altitudes, 900 m., 2,700 m. and 3,600 m. (páramo), in order to preserve representative areas of the whole of the Andean range in this part of Colombia. The Garden signed a collaboration agreement with the University of Quindío and with other public and private institutions. At present scientific and planning advice is required to develop the Garden's facilities and satellite gardens.

TROPICAL BOTANIC GARDENS: THEIR ROLE
IN CONVERVATION AND DEVELOPMENT
ISBN 0–12–346850–7

History

In order to spread the ecological knowledge of the area, a group of people interested in the conservation of the regional flora of Quindío set up a non-profit making scientific, and cultural organization in 1978 called the Botanic Garden of Quindío. The Government of Quindío granted it official status through a resolution (No. OJ 037) which was approved on the 5th of June of the following year.

Later, the Local Autonomous Corporation of Quindío, the official institution in charge of environmental conservation in the region, granted eight hectares of natural forest on loan to the Garden, land which belongs to the Forestry Reserve of Bremen. A further two adjacent hectares were given by the Sociedad Bosquinsa S.A. Since then important activities have been carried out in the area. The Garden is situated in the town of Filandia, at 1,950 m., and is 12 km. by road from the city of Armenia. Its location in the hydrographic basin of the river Roble, which supplies water to the reservoir of the town of Montenegro, is exceptional not only due to the high biological diversity of the area, but also for its excellent possibilities for protection and management. The National Federation of Coffee Producers, through the Department of the Quindío Committee is also linked to the conservation and management project of the basin.

The planning of the Garden

In 1984 the Garden signed an agreement with the following institutions: the Government of Quindío, The Local Autonomous Corporation of Quindío, the Committee of Coffee Producers of Quindío, the University of Quindío and the local horticultural club. In this way all these institutions were integrated in the common task of developing a series of research and conservation projects to protect the fast disappearing natural diversity of the region. This diversity is of great national and international scientific interest due to its special geographical location, ecology and floristic diversity, although these have been overlooked until now.

In the following year, 1985, thanks to a grant from Colciencias and the Second Botanical Expedition Foundation, the Botanic Garden of Quindío signed a contract with Arthur Simon Company Ltd for their specialized environmental planning services. This firm had to submit a preliminary study on the possibilities of reorganizing the Garden, and its impact on the agricultural activities and ecological requirements of the region of Quindío. As a result of a serious and very complete evaluation, the contractor produced a document entitled "Caracterización ecológica regional y anteproyecto del Jardín Botánico El Roble (Quindío)" [Ecological characterization of the region and draft project of the Botanic Garden El Roble (Quindío)]. The parameters required for the running and development of the Botanic Garden are given in the draft, and it also established that conservation would be the primary aim of the Garden, although not to the exclusion of other scientific research and educational activities. It noted that with the alarming state of disturbance of ecological systems in Quindío, it would be necessary to create a series of satellite gardens to represent the different thermic levels, ecosystems and geographic areas of the region, as essential natural reserves.

The Board of Trustees of the Garden adopted the recommendations outlined by this research, and declared that the activities and resources of the Garden should focus on the study of applied ecology. That is, it should favour practical projects with possible benefits for the "campesinos" and farmers of Quindío, stressing also the environmental education of the citizens. While developing these suggestions, we are confident that soon the Garden will have five protected and managed stations in five different thermic zones of Quindío: Páramo (3,600 m.); Salento (2,700 m.); Filandia (1,950 m.); Calarcá (1,490 m.), and Rio La Vieja (900 m.).

Institutions contributing to the Botanic Garden of Quindío

a) The University of Quindío took over the scientific organization of the Garden with the help of a biologist, Dr Cristina Vélez Nauer. In exchange, the Garden provides scientific information for students and teachers at the university. The Garden has been supporting the university as part of a project entitled "Hacia un estudio de la Flora del Departamento del Quindío" (Towards the study of the flora of the Department of Quindío). Due to the small size of the Department (1,950 sq. km.) there is little justification for establishing a separate herbarium for the Garden. Using a computer programme received as a gift from the Cartagena Botanic Garden, we are presently working on the development of software for herbarium management and we hope also that the computer will process information on lists of local species that the Garden has received from the Institute of Natural History of the National University.

b) The Bank of the Republic made a contribution to the Garden to contribute towards the costs of obtaining scientific and managerial advice from the Director of the Cartagena Botanic Garden, "Guillermo Piñeres". This botanic garden specializes in the study and micropropagation of economically important species, an important source of foreign currency through sales abroad. The Bank made a number of significant donations towards the establishment of a library for the Garden.

c) The Government of Quindío, through the Secretariat of Agriculture and Environment, employed a biologist, Arlés López, as Assistant Director of the Garden.

d) A private society, Bosquinsa S.A., as previously mentioned, presented the Garden with two hectares of natural forest on permanent loan.

e) An official institution known as the Local Autonomous Corporation of Quindío provided administrative and technical support for the Garden as well as eight hectares of the forest of Bremen.

f) A farmer, Fernando Uribe, offered ten additional hectares of páramo (at an altitude of 3,600 m.) in the town of Génova on permanent loan.

The project of Calarcá

a) Ecological justification: The Andes mountain range, and particularly the Central Cordillera, is geologically very young. Through colonization, the population established itself between the altitudinal range of 1,000-2,000 m. due to the mildness of the climate and the richness of useful plant species there to man. This resulted in the total removal of forest which, as it known today, is very important for climatic regulation, a factor which was ignored in the Andes due to high population pressure. The country still has a few areas of this forest type, in which conservation, propagation, and plant protection become more urgent and essential daily.

b) Location: It is situated on the outskirts of the city of Calarcá, Quindío, at 1,490 m. altitude and its survival until now was nothing short of miraculous. It is over 14 hectares in size and is located at the coordinates N4°31' 06" and E75°47'. It has a mean temperature of 19°C.

c) Ecological evaluation: After several visits the ecological value of this area became apparent due to the destruction of most similar areas not only in the Quindío area but also in the rest of Colombia. The high value and potential of the area is due to the rich diversity of the flora of the region, especially of that represented in the vegetation belt now largely removed for coffee plantations, and the development of sciophilous forest ground cover with an excellent regenerative capacity.

Amongst the economically important floristic elements in it are *Persea carulea* (little avocado), *Ocotea* sp. and *Nectandra* sp. (laurel), *Cecropia peltata* (yarumo), *Ormosea* sp. (chocho), *Ochroma lagopus* (balso), *Juglans neotropica* (black cedar), *Guadua angustifolia* (gudual) and *Anacardium excelsum* (Caracoli). A diverse and varied flora of shrubby species, climbers, creepers, lichens, mosses, ferns, bromeliads, etc. can also be found associated with the forest tree species. The forest also has a rich fauna, especially birds, and considerable scenic beauty Situated close to an urban centre it also has potential for wider use for educational, ecological and cultural activities and a relict example of the typical natural forest of the region at this altitude.

d) Protection and management: The forest has been monitored and protected very closely to avoid damage. This work has been carried out by the Natural Resources Section of the Local Autonomous Corporation of Quindío. In 1988 the Garden approved a resolution to find the finance necessary to buy the locality, which is expensive (US $155,000) due to its proximity to the city. Finally, in March 1989 five contributors, three of them from the public sector, were found. Further to this, it will be necessary to raise money for the development of the area (US$190,000 approximately), and to buy research equipment and develop new facilities, such as laboratories etc. (US$30,000).

The main offices of the Garden will be based at the station of Calarcá, and other satellite gardens will be established at different thermic zones. In addition, greenhouses, nurseries, propagation units, paths, gardens, education and other public services will be constructed at this station.

Due to the biological richness of the neotropical flora in Colombia, and particularly of the flora of Quindío, it is planned to develop an environmental education programme about the flora of the region using the resources of the five stations of the Botanic Garden.

Measures that are required to help conserve the flora of Mauritius and Rodrigues in *ex situ* collections

A.W. Owadally, M.E. Dulloo

Forestry Department, Curepipe, Mauritius

and

Wendy Strahm

c/o Forestry Quarters, Black River, Mauritius

Summary

Mauritius and Rodrigues have two of the most threatened floras in the world. Over 150 species of plants are threatened with extinction, out of which at least 20 species are known from less than 10 individuals. Although the long term conservation of these species must be in the wild, many of the rarest species can only be saved from extinction by *ex situ* propagation. A list of these species, along with a few others requiring urgent action, has been drawn up. This paper lists the species that may most benefit from *ex situ* cultivation with the ultimate goal being their re-introduction into the wild and into botanic gardens.

Methods of propagation are briefly discussed. With regard to endangered species, non-destructive means of propagation should be used in preference to destructive ones. The facilities available in Mauritius and Rodrigues for undertaking propagation of rare plants are mentioned. In Mauritius, the Sir Seewoosagur Ramgoolam Botanic Garden at Pamplemousses and the Curepipe Botanic Garden have the potential for contributing much to the conservation of the Mascarene flora. Rodrigues does not have a Botanic Garden. It is proposed to establish a small

TROPICAL BOTANIC GARDENS: THEIR ROLE
IN CONVERVATION AND DEVELOPMENT
ISBN 0—12—346850—7

satellite botanic garden there for conservation, research and recreation. It will be difficult to have the entire flora of Mauritius and Rodrigues in botanic garden collections in adequate numbers. So for long-term conservation, species held must be carefully selected and propagated with a view to their eventual planting in the wild and their protection in nature reserves.

Propagation of species in *ex situ* collections is not enough. The plants must be well documented so that there is no doubt about their provenances and every effort must be made to avoid hybridization. Both of these are essential for their eventual re-introduction to the wild.

Returning plants to Mauritius after they have been propagated in overseas collections must also be improved. With good planning, the entire flora of Mauritius can be saved both in botanic garden collections and in the wild. However, the wild populations are being hampered by exotic weeds and every effort must be made to save the wild habitats from degradation.

Introduction

Mauritius and its island dependency Rodrigues lie in the Western Indian Ocean, some 900 and 1,500 km. respectively east of Madagascar. Both islands are of volcanic origin; Mauritius has been dated at 7.8 million years old (McDougall & Chamalaun, 1969) and Rodrigues at about 1.5 million years old (McDougall et al., 1965). Mauritius covers 1,865 km^2 with its highest peak reaching 828 m., whereas Rodrigues is 109 km^2 in extent, attaining 393 m. in height. Both islands are part of the Mascarene Islands which also include Réunion.

Mauritius and Rodrigues have an endemic flora of about 290 and 46 taxa of flowering plants, most of which are "Red Data Book" species. In accordance with IUCN guidelines (Lucas and Synge, 1978), about 94 (32%) endemic taxa from Mauritius (Strahm, in prep.) and 20 (43%) taxa from Rodrigues (Strahm, 1989) are endangered. There are at least 38 Mauritian endemic plant species and 9 from Rodrigues that are extinct.

Though these islands were only colonised and settled some three centuries ago, the impact of man on the flora and fauna of the islands has been tremendous (Cheke, 1987). Before the arrival of man, the islands were covered with evergreen forests and palm savannah, the latter which has disappeared, except for a relict found on Round Island. The native forests are restricted to mountain, river and nature reserves. On Mauritius, upland and lowland forests can still be distinguished, while on Rodrigues no intact native forest remains.

Native species are highly wind resistant, having evolved in the presence of regular cyclones. They are mostly slow-growing but are long-lived, and grow better in groups rather than as isolated trees. They do not readily stand the effects of browsing and cutting as they have very few dormant buds. They compete poorly with gregarious exotic species.

What remains of the native forest is threatened by introduced plants and animals, and illegal cutting for firewood (Owadally, 1980; Cheke, 1987; Strahm, 1989). Exotic and invasive species such as *Psidium cattleianum* Sabine, *Ligustrum robustum* Blume var. *walkeri* (Decaisne) Mansf., *Rubus alceifolius* Poiret, *Ravenala madagascariensis* Sonnerat, *Homalanthus populifolius* Graham, *Hiptage benghalensis* (L.) Kurz, *Schinus terebinthifolius* Raddi, to name only a few, strongly compete with the native flora. Introduced animals, such as Russa deer *Cervus timorensis*, pigs *Sus scrofa*, Javanese Macaque monkeys *Macaca fascicularis*, and Black Rats *Rattus rattus* eat or otherwise damage saplings, flowers, fruits and seeds. Finally, there is illegal cutting for firewood, although this mainly affects exotic forest plantations.

Since 1951, several nature reserves have been proclaimed by law to protect different vegetation types. 2.4% of the area of Mauritius is included within these nature reserves, although less than 1% of Mauritius is under some sort of native forest. In 1983 several groups of plants, such as all orchids, ferns, ebonies and trochetias, were protected. This protection had to be modified as it was considered by the Court to be unconstitutional and in 1986 it was amended to include only plants found on Crown Lands, thus reducing their protection.

In situ plant conservation has to be promoted, as indigenous resources will continue to play an important rôle in the genetic development of the species far into the future (Barnes, 1985). Our chief aim is to conserve plants *in situ* as well as possible, but the situation is so serious that this is not enough. In addition, *ex situ* conservation has to be used to protect and multiply rare plants if they are to be saved from extinction.

Plant propagation

For a number of years, the Forestry Service has been propagating some indigenous plant species from seeds and cuttings. In the beginning, emphasis was not given to rare plants, but on any plant for which seeds were readily available. We have found that for a plant to succeed on a large scale in *ex situ* conservation, it must have some ornamental, medicinal or other interest.

Some of the plants propagated have even been used for enrichment planting in nature reserves (Owadally, 1972; 1973). Over the last fifteen years or so, more and more efforts have been concentrated on the propagation of endangered plants. Species such as *Gaertnera longifolia, Canarium paniculatum, Hyophorbe vaughanii, Tambourissa peltata, T. amplifolia, Myonima violacea, Trochetia boutoniana, Coffea mauritiana, Gaertnera psychotrioides, Bertiera zaluzania, Pandanus pyramidalis, Tetrataxis salicifolia, Calophyllum eputamen, Diospyros boutoniana, Cylindrocline commersonii, Diospyros pterocalyx, Diospyros neraudii* and *Tectiphiala ferox* have been propagated and their provenances are noted in the seed register book kept at the headquarters of the Forestry Service. The following information is recorded: date of collection, site, data on sowing, date of germination, number of plants successfully raised and sites to where the seedlings have been finally transplanted.

From June-September 1985 the Dublin/Cambridge Expedition, headed by P. Wyse Jackson, came to Mauritius for plant conservation work. They helped us set up a small mist propagation bench with heating cables for rooting cuttings, with a fair amount of success (Wyse Jackson et. al., 1988). The Forestry Service has now built a concrete greenhouse for the propagation of rare plants. This is now being equipped with automatic mist sprayers and heated benches.

The Agricultural Services at Barkly Nursery, Beau Bassin, are setting up a modern tissue culture laboratory which can also be used to propagate rare plants.

The Forestry Service has had immense difficulties in receiving plants propagated and cultivated abroad, such as *Begonia salaziensis*, *Gastonia rodriguesiana* and *Dombeya rodriguesiana* from Kew. Some of these plants which were raised at great cost and care, were lost due to long delays in air freight and processing by custom and quarantine authorities, as well as damage caused by the application of insecticides and fungicides.

Plant reintroduction

Once the rare plants have been produced in nurseries and laboratories, they must be planted in sites with security of tenure. In the past, the Forestry Service has used private gardens. Unfortunately, after the death of the owner, the protection of the plants can no longer be guaranteed.

Public compounds have also been used to grow rare plants. Although this is safer, it still poses many problems, such as the erection of new buildings in the compound, accidental trampling by the public, vandalism, etc.

Both of the botanic gardens, at Curepipe and Pamplemousses have been used as sites for planting indigenous plants of Mauritius and Rodrigues. At Curepipe, the Forestry Service, with the collaboration of the Municipality of Curepipe, has fenced about one hectare of land (Owadally, 1979). A collection of native trees and shrubs was planted and several species of *Pandanus* (*P. carmichaelii, P. glaucocephalus, P. palustris*) have been introduced in a marshy area near the garden's pond. Progress has not been as good as expected as the plants seem to suffer from exposure. Rapidly growing indigenous "nurse" species have been planted to shade and to improve the micro-climate for the rare plants.

At the Sir Seewoosagur Ramgoolam Botanic Gardens, Pamplemousses, where a larger site is available, a new section for Mascarene plants has been laid out very successfully. From 1980 to 1988, the number of native plants in the Gardens has doubled from 25 to 50. Unfortunately, due to unforeseen circumstances, the plants may now have to be removed and transplanted to another part of the Gardens. One of the long-term policies of the Pamplemousses Gardens should be their full involvement in conservation of endangered species.

The creation of arboreta for indigenous plants both in Mauritius and Rodrigues must be undertaken. These need to be established permanently in various localities with different climates, altitudes and soil types. As no public gardens exist in Rodrigues, it is essential that an arboretum or botanic garden be set up without undue delay.

As long as the plants produced *ex situ* are pure and their provenances well known, they can also be used for enrichment planting in nature reserves which are far better protected in terms of land tenure. There may also be less risk of future hybridization with related species.

Some case histories

Although we cannot dwell at length here on all the species for which propagation and cultivation have been attempted, some examples will be used to raise and illustrate important issues concerning *ex situ* conservation in Mauritius.

Hyophorbe amaricaulis

The struggle to save the sole surviving individual of the palm species *Hyophorbe amaricaulis* in the Curepipe Botanic Garden has been going on for several years. In 1979, a complete inflorescence was sent by air, through diplomatic bag, to Lucknow National Botanic Garden, India, for tissue culture (Owadally, 1979). Unfortunately, it did not arrive in a viable state. Seeds have also been collected and sown with various treatments, but they all failed to germinate. In 1985, unripe seeds were sent to Trinity College, Dublin, and Wye College, England for micropropagation. Both succeeded in germinating the embryonic tissue and getting shoot and root formation, which failed later (Douglas, 1987).

On May 11, 1988, an inflorescence was sent to the Barkly tissue culture laboratory, Mauritius, with no positive results. On June 9, 1988, one inflorescence opened and the pollen was collected, as the male flowers open before the female. The pollen was kept in the refrigerator under dry conditions after being dehydrated at 40°C in an oven. After one month the female flowers began to open, which were gradually hand-pollinated and bagged. A number set seed and these have been treated with giberellic acid, and the tree is treated with a slow release fertilizer. As at April 1989, there are 25 ripening fruits on the tree, which are still far from being ripe. Although the palm produces flowers regularly, unfortunately, the inflorescences generally drop off before any pollination can be attempted.

Tectiphiala ferox

This monotypic palm was not described until 1978. It is known from only a few trees and ripe seeds are difficult to obtain. In 1980, seeds were collected from Crown Land Gouly Père and sown in a mixture of soil and pine sawdust in the proportions 2:1. After 40 days a few seeds germinated, and were eventually planted out. Individuals planted in the open did not survive but those planted in shady areas did.

Crinum mauritianum

Only one population of this species is left occurring in the Midlands area. The population was threatened as its site was due to be flooded as a reservoir. Fortunately, the dam was never completed and this population still survives. Because it is very ornamental and grows readily from seed, it has been grown in private gardens and it is also grown in botanic gardens throughout the world. The survival of this species is ensured due to *ex situ* cultivation, but it is still endangered in the wild.

Sideroxylon grandiflorum

The "tambalacoque", *Sideroxylon grandiflorum*, a tall upland forest tree, is rare and does not regenerate readily in the wild. Germination is low and seeds are difficult to obtain because monkeys destroy most of the fruit. A collection of 80 seeds was collected on December 3, 1971, from a tree found at Gorette by the Plaine Champagne/Chamarel road. On December 30, 1971, these seeds were sown in humus rich brown forest soil. Two of them germinated on April 4, 1972. On September 8, 1973, the only seedling that survived was planted at the campus of the University of Mauritius. This fine plant is flourishing and is now over five m. tall. Unfortunately, the parent tree fell down on October 10, 1983, as the base was rotten due to damage caused to the tree when the road was tarred in 1972. Propagation of this species remains a problem.

Hibiscus spp.

All of the five indigenous *Hibiscus* species found on Mauritius and Rodrigues are threatened. Fortunately, they have been propagated easily by seeds and cuttings and are grown in several gardens. Unfortunately, they may all hybridize, thus extreme care must be taken to ensure that the cultivated plants are pure.

Hibiscus liliiflorus, endemic to Rodrigues, was known from a single individual, which unfortunately was killed by people taking cuttings for propagation, as well as by goats browsing (Strahm, 1982). Only one cutting of this plant is known to have taken, and is grown in the garden of a Catholic brother. A cutting was grafted onto *H. rosa-sinensis* stock in Pamplemousses Gardens. This plant produces seeds each year but they may be hybrids with other *Hibiscus* spp. growing nearby. Fortunately, two more wild individuals have since been discovered on Rodrigues, and plants have been propagated from the seeds of these individuals.

Coffea spp.

Coffea is one of the few genera on Mauritius which may have an easily apparent economic importance. There are two endemic species, *Coffea macrocarpa* and *C. myrtifolia* and a third, *C. mauritiana*, endemic to Mauritius and Réunion. The only species really endangered is *C. myrtifolia*, which was only known from a small population in the Magenta forest. This species has been grown from cuttings and several individuals are now in cultivation.

Ficus laterifolia

This species in only known from three trees growing along river banks, one of which is behind the Curepipe Forestry nursery. Since the tree is easily accessible, it has been multiplied by layering and cuttings and the progeny planted elsewhere along river banks.

Pandanus pyramidalis

This species is only known from a single male tree growing in a vulnerable site. One branch produced roots from an air layering, but unfortunately the compost around the branch was removed by vandals. The tree is in a very poor state.

Ramosmania (Randia) heterophylla

This member of the Rubiaceae was thought to be extinct on Rodrigues, but was re-discovered in 1981 by a secondary school teacher working on the island. The tree was considered by Rodriguans to have medicinal properties and was constantly being cut by them. To protect it, the plant was then progressively surrounded by four fences, but still it was not safe, even though a permanent guard watched it.

In 1986, a rescue operation was mounted and two cuttings were sent to the Royal Botanic Gardens, Kew for propagation. This needed the combined efforts of Rodriguans, several airlines, the Ministry of Agriculture and a few enthusiasts, both local and at Kew, to save this plant.

On 20 April, 1989 one young plant of *Ramosmania* was returned to Mauritius from Kew propagated from a young plant raised from one of the cuttings. It did not survive and the rapid and extensive propagation of *Ramosmania* remains a difficult problem. As well as that, its repatriation poses several logistic problems, such as national quarantine barriers, which must be solved if rare plants are to be raised in other countries and regularly returned to their country of origin.

Some of the notable rare endemic species (Appendix 3) which have been raised from cuttings (with various rooting hormone treatments) include *Embelia micrantha, Elaeocarpus bojeri, Hibiscus genevii, H. fragilis, Pisonia costata, Dombeya acutangula, Chasalia lanceolata, Vernonia populifolia, Poupartia borbonica* and a fairly remarkable amount of success has been registered. In others, the methods used so far have not proved successful. Many of the plants listed in Appendix 1. are so reduced in number that plant materials are not available to try different methods of propagation. Care must be taken not to destroy or eliminate the plant completely from the wild.

About 50 rare and endangered plant species from Mauritius and Rodrigues have been noted in cultivation in botanic gardens around the world outside of Mauritius (P. Wyse Jackson, pers. comm, 1987).

Important plants for cultivation

There are over 300 species endemic to Mauritius and Rodrigues, (plus a further 150 endemic to the Mascarene Islands and other rare native species). There is therefore, a large number of species in need of cultivation. A list of priorities must be established, and Appendix 1 lists the Mauritian endangered plants known from less than 20 individuals or only found in one or two localities in the wild. The list contains 121 taxa. This list includes some species which are already fairly common in cultivation, but which are still endangered in the wild. The palms in particular are widely cultivated. Both *Acanthophoenix rubra* and *Dictyosperma album* are commercially grown for their palm heart, eaten either cooked or raw. *Hyophorbe lagenicaulis*, the "Round Island bottle palm", is cultivated throughout the world, but only 7 adult trees remain on Round Island. *Latania loddigesii*, the "Round Island blue latan", is also extensively grown, but only two populations remain, on Round Island and on another island nearby, Gunner's Quoin, although the numbers are still high on the former island.

Crinum mauritianum, Hibiscus genevii and *Trochetia boutoniana* are grown in several private and public gardens, although more will need to be cultivated on Mauritius before they can be said to be out of danger.

Appendix 2. lists the endangered species of Rodrigues which are priorities for cultivation and need to be multiplied for eventual reintroduction planting.

Conclusions

No efforts should be spared, both in Mauritius and abroad, for the conservation of all our plants both *in situ* and *ex situ*. To help *ex situ* conservation and propagation, a thorough knowledge of the status of the plant in its natural habitat is essential, so as to avoid complete destruction of extremely rare plants. Detailed records of all information regarding the plants to be propagated, for example, their origin, method of propagation, sites (both local and abroad) where planted, and so on, must be kept. Once the plants have been propagated, they must be planted on sites which have a security of tenure. Botanic gardens are expected to play a key rôle in this connection. Moreover, a few arboreta must be created both in Mauritius and in Rodrigues to ensure the continuous maintenance and propagation of these plants. These must be replicated elsewhere to avoid total loss by natural calamities such as cyclones. In order to be able to use these plants for further propagation for enrichment planting in nature reserves, all practical measures must be taken to avoid hybridization. In some extreme cases, propagation may have to be carried out abroad. This entails close collaboration and easier exchange of plant material between several authorities and organizations in different parts of the world. The phytosanitary procedures for the exchange of rare plants to be raised or already raised *ex situ* must be reviewed in order not to waste time and effort and to ensure that these plants reach their destination in the best possible condition. Last but not least, research and training in the field of rare plant propagation must be maintained.

Acknowledgements

We would like to thank the Ministry of Agriculture, Fisheries and Natural Resources for encouraging this work and the staff of the Forestry Service for their help and support. Since 1983, we have received funding from the World Wide Fund for Nature (WWF) which has provided a grant for WAS. More recently the Australian High Commission has provided funds for a greenhouse and WWF have provided propagation equipment. We would also like to thank the Mauritius Herbarium and many local botanists for help and encouragement.

References

Barnes, R.D. (1985). The importance of the natural forests in breeding tropical pines. Paper presented at International Workshop about the state of the art in forest genetics and tree breeding. IX World Forestry Congress, Mexico City, 1st-10th July, 1985. 8 pp.

Cheke, A. (1987). An ecological history of the Mascarene Islands, with particular reference to extinctions and introductions of land vertebrates, p. 5-89. In: *Studies of Mascarene Island Birds* (Ed. A.W. Diamond). Cambridge University Press, Cambridge. 458 pp.

Douglas, G.C. (1987). Embryo culture of a rare plant, *Hyophorbe amaricaulis* (Mauritius) (Palmae). *J. Plant Physiology, 130*, 73-77.

Lucas, G. and Synge, H. (1978). *The IUCN Plant Red Data Book.* IUCN, Switzerland, 540 pp.

McDougall et al. (1965). A geological reconnaissance of Rodrigues Island, Indian Ocean. *Nature, 206*, 26-27.

McDougall & Chamalaun (1969). Isotopic dating and geomagnetic polarity studies on volcanic rocks from Mauritius, Indian Ocean. *Bull. Geol. Soc. Am., 80*, 1419-1442.

Owadally, A.W. (1972). *Annual Report of the Forestry Service for the year 1972,* Mauritius No. 21 of 1974, Government Printers, Port Louis, Mauritius.

Owadally, A.W. (1973). Les fôrets naturelles de l'Ile Maurice. *Info-Nature Ile Réunion*, No. Spécial hors serie, "La fôret", 88-94.

Owadally, A.W. (1980). Some forest pests and diseases in Mauritius. *Rev. Agric. Sucrière. Ile Maurice*, 59, 76-94.

Strahm, W. (1989). *Plant Red Data Book for Rodrigues.* Koeltz Scientific Books, Konigstein. 241 pp.

Wyse Jackson, P.S., Strahm, W., Cronk, Q.C.B. and Parnell, J.A.N. (1988). The propagation of endangered plants in Mauritius. *Moorea,* 7, 35-45.

Appendix 1.
Endangered plants of Mauritius known from less than 20 wild individuals or 1-2 localities.

Although there are other endangered and vulnerable species which urgently need to be cultivated but not included in this list, nevertheless, emphasis should be placed on the most critically endangered species listed here. Species marked with an asterisk are already in cultivation in more than one permanent collection.

Family / Species	Status
Amarantaceae	
Aerva congesta Balf. f.	Mauritius: E Rodrigues: Ex
Amaryllidaceae	
**Crinum mauritianum* Lodd.	Mauritius: E
Anacardiaceae	
Poupartia borbonica Gmel.	Mauritius: E Réunion: ?
Annonaceae	
Xylopia amplexicaulis (Lam.) Baillon	Mauritius: E
Apocynaceae	
Carissa xylopicron Thouars	Mauritius: E Rodrigues: V Réunion: Ex
Ochrosia borbonica J.F. Gmelin	Mauritius: E Réunion: ?
Araliaceae	
Polyscias gracilis Marais	Mauritius: E
Polyscias neraudiana (Drake) R. Viguier	Mauritius: E
Asclepiadaceae	
Secamone volubilis (Lam.) Marais	Mauritius: E Réunion: ?
Begoniaceae	
**Begonia salaziensis* (Gaudich.) Warb.	Mauritius: E Réunion: ?
Campanulaceae	
Lobelia serpens Lam. var. *serpens*	Mauritius: E Réunion: ?
Lobelia serpens Lam. var. *cheiranthifolia* (Pres.) DC.	Mauritius: E Madagascar: ?
Lobelia serpens Lam. var. *puberula* E. Wimm.	Mauritius: E

Taxon	Distribution
**Nesocodon mauritianus* (I.K.B. Richardson) Thulin	Mauritius: E
Compositae	
**Cylindrocline commersoni* Cass.	Mauritius: E
Faujasia pinifolia Cass.	Mauritius: E Réunion: ?
Faujasia reticulata Benth.	Mauritius: E
Senecio appendiculatus DC.	Mauritius: E/Ex
Ebenaceae	
Diospyros angulata Poiret	Mauritius: E
**Diospyros egrettarum* I.B.K. Richardson	Mauritius: E
Elaeocarpaceae	
Elaeocarpus bojeri R.E. Vaughan	Mauritius: E
Elaeocarpus integrifolius Lam.	Mauritius: E
Euphorbiaceae	
Acalypha integrifolia Willd. subsp. *integrifolia* var. *longifolia* (Müll.Arg.) Coode	Mauritius: E
Acalypha integrifolia Willd. subsp. *marginata* (Poiret) Coode var. *marginata*	Mauritius: E
Acalypha integrifolia Willd. subsp. *integrifolia* var. *parvifolia* (Baillon ex Müll. Arg.) Pax & Hoffm.	Mauritius: E
Claoxylon linostachys Baillon subsp. *linostachys*	Mauritius: E
Claoxylon linostachys Baillon subsp. *brachyphyllum* (Croizat) Coode	Mauritius: E
Claoxylon linostachys Baillon subsp. *pedicellare* Coode	Mauritius: E
Croton vaughanii Croizat	Mauritius: E
Drypetes caustica (Frappier ex Cordem.)	Mauritius: E Réunion: ?
Phyllanthus oppositifolius Baillon Müll. Arg.	Mauritius: E
Phyllanthus phillyreifolis Poiret var. *stylifer* Coode	Mauritius: E
Phyllanthus phillyreifolis Poiret var. *gracilipes* Coode	Mauritius: E
Phyllanthus phillyreifolis Poiret var. *triangularis* Müll. Arg.	Mauritius: E
Phyllanthus pileostigma Coode	Mauritius: E
Phyllanthus revaughanii Coode	Mauritius: E
Flacourtiaceae	
Casearia mauritiana Bosser	Mauritius: E

Taxon	Distribution
Casearia tinifolia Vent.	Mauritius: E/Ex
Lauraceae	
Ocotea lancilimba Kosterm.	Mauritius: E
Leguminosae	
Albizia vaughanii Brenan	Mauritius: E
Gagnebina pterocarpa (Lam.) Baillon	Mauritius: E Madagascar: ?
Zornia vaughaniana Mohlenbr.	Mauritius: E
Liliaceae	
Astelia hemichrysa (Lam.) Kunth	Mauritius: E World: Nt
**Dracaena concinna* Kunth	Mauritius: E
Dracaena floribunda Baker	Mauritius: E
*Lomatophyllum tormentorii Marais	Mauritius: E
Korthalsella opuntia (Thunb.) Merrill var. *bojeri* (Van Tieghem) Danser	Mauritius: E Rodrigues: Ex Réunion: ?
Lythraceae	
**Tetrataxis salicifolia* (Thouars ex Tul.) Baker	Mauritius: E
Malvaceae	
**Hibiscus boryanus* DC.	Mauritius: E Réunion: ?
**Hibiscus fragilis* DC.	Mauritius: E
**Hibiscus genevii* Bojer ex Hook.	Mauritius
Melastomataceae	
Memecylon cordatum Lam.	Mauritius: E Réunion: ?
Meliaceae	
Turraea decandra (Cav.) Harms	Mauritius: E Réunion: ?
Turraea ovata (Cav.) Harms	Mauritius: E Réunion: ?
Turraea trichopoda (Baillon) Harms	Mauritius: E Réunion: ?
Monimiaceae	
Monimia ovalifolia Thouars	Mauritius: E Réunion: Nt
Tambourissa cocottensis D. Lorence	Mauritius: E
Tambourissa pedicellata Baker	Mauritius: E
Tambourissa tetragona (Boiv. ex Tul.) DC.	Mauritius: E
Moraceae	
**Ficus laterifolia* Vahl	Mauritius: E Réunion: ?

Myrsinaceae	
Badula crassa DC.	Mauritius: E Réunion: Ex
Badula platyphilla (DC.) Coode	Mauritius: E/Ex
Badula reticulata DC.	Mauritius: E/Ex
Embelia micrantha (DC.) DC.	Mauritius: E Réunion: ?
Myrtaceae	
Eugenia crassipetala Guého & A.J. Scott	Mauritius: E
Syzygium populifolium (Baker) Guého & A.J. Scott	Mauritius: E
Oleaceae	
Olax psittacorum (Lam.) Vahl	Mauritius: E Réunion: ?
Chionanthus ayresii A.J. Scott	Mauritius: E
Chionanthus boutonii A.J. Scott	Mauritius: E
Chionanthus broomeana (Horne ex Olivier) A.J. Scott subsp. *broomeana*	Mauritius: E
Orchidaceae	
Angraecum eburnea Thouars	Mauritius: E World: Nt
Angraecum [saccolabium]	Mauritius: E World: ?
Arnottia mauritiana Rich.	Mauritius: E Réunion: ?
Beclardia macrostachya (Thou.) A. Rich.	Mauritius: E Réunion: ?
Bulbophyllum incurvum Thouars	Mauritius: E World: ?
Bulbophyllum longiflorum Thouars	Mauritius: E World: ?
Eulophidium monophyllum (Rich.) Schl.	Mauritius: E
Goodyera nuda Thouars	Mauritius: E
Graphorkis scripta Lindl.	Mauritius: E World: ?
Liparis purpurascens Lindl.	Mauritius: E Réunion: ?
**Oeoniella aphrodite* (Balf. & S. Moore) Schlt.	Mauritius: E World: ?
Phajus longibracteatus Frapp.	Mauritius: E Réunion: ?
Phajus pulchellus Kranzl.	Mauritius: E
Phajus villosus Rchb. f.	Mauritius: E

Polystachya cultriformis (Thou.) Lindl. ex Spreng.	Mauritius: E World: ?
Solenangis aphylla (Thou.) Summerhayes	Mauritius: E World: ?
Palmae	
*_Acanthophoenix rubra_ (Bory) H. Wendl	Mauritius: E
*_Dictyosperma album_ (Bory) H. Wendl & Drude ex Scheff. var. *conjugatum* H.E. Moore & L.J. Guého	Mauritius: E
*_Dictyosperma album_ (Bory) H. Wendl & Drude ex Scheff. var.*album*	Mauritius: E Réunion: ?
Hyophorbe amaricaulis Mart.	Mauritius: E
*_Hyophorbe lagenicaulis_ (L.H. Bailey) H.E. Moore	Mauritius: E
Hyophorbe vaughanii L.H. Bailey	Mauritius: E
Latania loddigesii Mart.	Mauritius: E
*_Tectiphiala ferox_ H.E. Moore	Mauritius: E
Pandanaceae	
Pandanus carmichaelii Vaughan & Wiehe	Mauritius: E
Pandanus glaucocephalus Vaughan & Wiehe	Mauritius: E
*_Pandanus microcarpus_ Balf. f.	Mauritius: E
Pandanus palustris Thouars	Mauritius: E
Pandanus pyramidalis Barkly ex Balf.f.	Mauritius: E
Rhamnaceae	
Glouania tiliifolia	Mauritius: E Réunion: ?
Rubiaceae	
Chassalia boryana DC.	Mauritius: E
Chassalia grandifolia DC.	Mauritius: E/Ex
Chassalia lanceolata (Poiret) A. Chev. subsp. *latifolia* Verdc.	Mauritius: E
Chassalia lanceolata (Poiret) A. Chev. subsp. *lanceolata*	Mauritius: E
Gaertnera hirtiflora Verdc.	Mauritius: E
Gaertnera longifolia Bojer var. *longifolia*	Mauritius: E
Gaertnera pendula Bojer	Mauritius: E
Gaertnera sp. A.	Mauritius: E
Gaertnera truncata A.DC.	Mauritius: E
Rutaceae	
Euodia chapelieri Baillon var. *sessilis* Coode	Mauritius: E
Euodia obtusifolia DC. subsp. *gigas* Vaughan ex Coode var. *brachypoda* Coode	Mauritius: E

Zanthoxylum heterophyllum (Lam.) Smith	Mauritius: E Rodrigues: Ex Réunion: ?
Sterculiaceae	
**Dombeya mauritiana* Friedmann	Mauritius: E
Dombeya populnea (Cav.) Baker	Mauritius: E Réunion: E
**Trochetia boutoniana* Friedmann	Mauritius: E
Urticaceae	
**Pilea articulata* Wedd.	Mauritius: E
Pilea boehmerioides Wedd.	Mauritius: E
Pilea cataractae Marais	Mauritius: E
Pilea cocottei Marais	Mauritius: E
Pilea pollicaris Marais	Mauritius: E

Key: Nt - not threatened, V - vulnerable, E - endangered, Ex - extinct and ? - unknown status.

Appendix 2.
Endangered plants of Rodrigues in need of cultivation

Family / Species	Status
Anacardiaceae	
Poupartia castanea (Baker) Engl.	Rodrigues: E
Araliaceae	
Gastonia rodriguesiana Marais	Rodrigues: E
Campanulaceae	
Lobelia vagans Balf. f.	Rodrigues: E
Euphorbiaceae	
Phyllanthus casticum Willemet f.	Rodrigues: E World: I
Securinega durissima J. Gmelin	Rodrigues: E Mauritius: Nt Réunion: ?
Flacourtiaceae	
Scolopia heterophylla (Lam.) Sleumer	Rodrigues: E Mauritius: Nt Réunion: E
Flagellariaceae	
Flagellaria indica L.	Rodrigues: E World: Nt
Lecythidaceae	
Foetidia rodriguesiana F. Friedmann	Rodrigues: E
Liliaceae	
Lomatophyllum lomatophylloides (Balf. f.) Marais	Rodrigues: E
Malvaceae	
Hibiscus liliiflorus Cav.	Rodrigues: E Réunion: Ex
Meliaceae	
Turraea laciniata (Balf. f.) Harms	Rodrigues: E
Myoporaceae	
Myoporum mauritianum A. DC.	Rodrigues: E Mauritius: Ex
Myrsinaceae	
Badula balfouriana (O.Kuntze) Mez.	Rodrigues: E
Myrtaceae	
Eugenia rodriguesensis Guého & A.J. Scott	Rodrigues: E
Nesogenaceae	
Nesogenes decumbens Balf. f.	Rodrigues: E/Ex
Orchidaceae	
Aeranthes arachnites Lindley var. *balfourii* S. Moore	Rodrigues: E

Jumellea recta (Thouars) Schltr.	Rodrigues: E World: ?
Oeoniella aphrodite (Balf. f. & S. Moore) Schltr.	Rodrigues: E World: ?
Palmae	
Dictyosperma album (Bory) H.A. Wendl. & Drude ex Scheffer var. *aureum* Balf. f.	Rodrigues: E
Hyophorbe verschaffeltii H.A. Wendl.	Rodrigues: E
Pittosporaceae	
Pittosporum balfourii Cuf.	Rodrigues: E
Rhamnaceae	
Gouania leguatii Guého	Rodrigues: E
Rubiaceae	
Antirhea bifurcata (Desr.) Hook. f.	Rodrigues: E Mauritius: Nt
Ramosmania heterophylla (Balf. f.) Tirvengadum & Verdcourt	Rodrigues: E
Scyphochlamys revoluta Balf. f.	Rodrigues: E
Rutaceae	
Vepris lanceolata (Lam.) G. Don	Rodrigues: E World: Nt
Zanthoxylum paniculatum Balf. f.	Rodrigues: E
Sapindaceae	
Allophyllus sp.	Rodrigues: E World: Nt
Dodonaea angustifolia L.f.	Rodrigues: E World: Nt
Sapotaceae	
Sideroxylon galeatum (A.W. Hill) Baehni	Rodrigues: E
Sterculiaceae	
Dombeya acutangula Cav.	Rodrigues: E Mauritius: R Réunion: ?
Dombeya rodriguesiana F. Friedmann	Rodrigues: E
Urticaceae	
Obetia ficifolia (Poiret) Gaudich.	Rodrigues: E Mauritius: Ex Réunion: E
Pilea balfourii Baker	Rodrigues: E

Key: Nt - not threatened, I - indeterminate, E - endangered, Ex - extinct and ? - unknown status.

Appendix 3
Plant Propagation: Percentage rooting of cuttings using different rooting hormone

		...SYNF*		SYN 1:1*		SER 1*		SER 2*		SER 3*	
	Species	n	%r +	n	%r	n	%r	n	%r	n	%r
1.	*Olax psittacorum*	45	0								
2.	*Monimia ovalifolia*	13	0	6	33					5	0
3.	*Drypetes caustica*	10	0	34	5.8			8	0	10	0
4.	*Coffea myrtifolia*	72	72	11	18						
5.	*Trochetia trifolia*			9	0			12	0		
6.	*Diospyros hemiteles*	10	0	18	0			18		20	
7.	*Tetrataxis salicifolia*	50	16	83	0			13	15		
8.	*Weinmania tinctoria*			7	0			4	0		
9.	*Erythrospermum monticolum*			2	0						
10.	*Claoxylon linostachys* var. *linostachys*			15	7	1	0			3	33
11.	*Embelia micrantha*			6	67			9	67		
12.	*Roussea simplex*			4	0	9	0				
13.	*Elaeocarpus bojeri*	13	31	5	80					2	50
14.	*Claoxylon linostachys* var. *brachyphyllum*			19	26						
15.	*Euodia obtusifolia*	3	0	7	0					4	0
16.	*Chionanthus boutonii*			10	0	7	43				
17.	*Polyscias gracilis*			6	17					6	0
18.	*Xylopia amplexicaulis*			10	0					4	0
19.	*Croton vaughanii*	15	0	20	20			5	20		
20.	*Ficus laterifolia*			5	20						
21.	*Helichrysum yuccaefolium*							3	0		
22.	*Trochetia blackburniana*							4	0		
23.	*Trochetia boutoniana*	6	50								
24.	*Eugenia lucida*	12	0								
25.	*Diospyros egrettarum*	11	64								

Species	...SYNF* n	%r+	SYN 1:1* n	%r	SER 1* n	%r	SER 2* n	%r	SER 3* n	%r
26. *Sideroxylon boutonianum*	10	?								
27. *Nesocodon mauritianum*			2	0						
28. *Dombeya mauritiana*	4	0			8	25				
29. *Tambourissa tau*	1	100								
30. *Hibiscus boryanus*			50	18					11	18
31. *Hibiscus genevii*	8	100	41	58						
32. *Hibiscus fragilis*	16	94	28	79						
33. *Pisonia costata*			7	43						
34. *Dombeya acutangula*			5	100						
35. *Psiadia trinerva*	2	50	4	75			3	67		
36. *Vernonia populifolia*	20	65	3	34			8			
37. *Cylindrocline commersonii*	2	0								
38. *Turraea casimiriana*									28	21
39. *Tarenna borbonica*									14	0
40. *Turraea rigida*	3	34	5	20			7	29		
41. *Badula multiflora*			2	0						
42. *Elaeocarpus integrifolius*			9	22				4	0	
43. *Chasalia lanceolata*	8	100	10	70						
44. *Ocotea mascarena*			3	0						
45. *Casearia mauritiana*	9	0	10	10						
46. *Chasalia capitata*	12	0	38	0						
47. *Foetidia mauritiana*	5	0	11	0						
48. *Trochetia uniflora*	13	15								
49. *Poupartia pubescens*							3	34		
50. *Poupartia borbonica*					3	34	6	67		
51. *Bakerelia hoyifolia*			6	0						
52. *Begonia salaziensis*					9	0				
53. *Xylopia lamarkii*									10	?
54. *Tambourissa pedicellata*			10	0					8	?
55. *Dombeya rodriguesiana*							3			
56. *Dombeya ferruginea*							6	33		

n = Nos. of cuttings taken
% r+ = percentage of cuttings which have rooted
Rooting Hormones*
SYNF = Synergol full strength
SYN 1:1 = Synergol to water in varying proportion
SER 1 = Seradix 1
SER 2 = Seradix 2
SER 3 = Seradix 3

Seradix – active constituent is 4–indol–zylbutyric acid

Synergol – contains potassium salt of indole butyric acid and the potassium salt of Naphthyl acetic acid (K-NAA) + trace elements and Boron; also fungicide.

The development of a university botanic garden in Brazil

Ariane Luna Peixoto

Universidade Federal Rural do Rio de Janeiro
Jardim Botânico, Brasil

Summary

The Universidade Federal rural do Rio de Janeiro is located in the Itaguai District, in the vicinity of Rio de Janeiro City, about 60 km from its main centre. The University organizes undergraduate courses in the areas of Forest Engineering, Veterinary, Agricultural Sciences, Domestic Economy, Biology, Chemistry and Mathematics. It also maintains a secondary level school in the Agricultural Science area. the University also offers graduate studies (Ph.D. and Masters) in such fields as Parasitology, Veterinary, Microbiology, Chemistry and Soil Physiology.

The campus buildings are in European style and built in the 40s; the area acts as an experimental station for many different fields of knowledge.

Eight years ago, as suggested and urged by the Biology Institute, and specially its Botany Department, an area of 18 ha was selected as the embryo of a University Botanic Garden. The main building of the botanic garden was one adapted from having been a resident professor's house. The occupation of the area followed the building of fences all around the area, road making, the creation of an artificial lake and general planning for the establishment of specimens.

Today the botanic garden has many problems as represented by its reduced staff of just 8 people recruited from other departments of the University, and most of them with little previous experience. The botanic garden is under the direct supervision of the University Dean and that allows it to be quite free from administrative bureaucracy. The

TROPICAL BOTANIC GARDENS: THEIR ROLE IN CONVERVATION AND DEVELOPMENT
ISBN 0–12–346850–7

Botany Department teachers are in charge of planning the work programme of the Garden and of keeping the herbarium.

In its first years the botanic garden's main interests have been: 1) the effective occupation of its area, which involves many problems, especially weed control; 2) the instruction and training of university students and teachers; 3) the collection and cultivation of specimens for the arboretum, especially those Brazilian species that are rare or endangered.

Students are trained in two priority tasks: collection, germination and seed cultivation of native species, mainly from the Atlantic Forest of Brazil, as well as learning about preservation measures in such ecosystems. Due to the wide field of study covered by its students the University Botanic Garden has proved to be quite an important source of information for those students, as well as a very important introductory training in general environmental management.

The Federal Rural University of Rio de Janeiro is situated in the municipality of Itaguai, 60 km from Rio, and offers higher courses in Agronomy, Forest Engineering, Veterinary and Agricultural Sciences, Domestic Economy, Biology, Chemistry, Physics and Mathematics. It has a technical agricultural college that trains professionals to intermediate level and higher courses, that is to say Masters and Doctorates in Veterinary Parasitology, Microbiology, Soil Physiology, Chemistry and Phyto-technology.

It is situated in an area of nearly 3 ha on which buildings and various experimental stations are variously located. The buildings of the university look European, and the organization of the courses follows the North American model. The green spaces of the campus are under the administration of public gardens and parks depeartment, which also maintains an orchard for raising plants to be used on the campus.

Nine years ago, at the request of professors of the Institute of Biology (in particular of the Department of Biology) an area of 18 hectares was reserved for the establishment of the botanic garden. In this area a house was adapted to serve as the head office, and plans were made for a road network, a lake and the general planning of the layout of a plant collection.

The running of the botanic garden is in the hands of a secretary, a typist, a shorthand typist, two gardeners, 3 agricultural labourers and a driver. The direction is provided by a departmental professor of botany who also undertakes a rôle in teaching and research in the department. The botanic garden itself is regarded as a department within the university with well defined goals and is allowed relative autonomy. It is linked administratively to the office of the Rector of the University, and the technical policy is provided by a team of botany department professors. In this way, the botanic garden can plan its cultural, scientific, educational and leisure activities with a high degree of independence.

During its first year of existance the botanic garden focussed its activities in three directions.

1. The use of the land within the garden.
2. Development of an orientation programme for visiting students and professors.
3. The collection and cultivation of rare native tree species, especially those in danger of extinction.

1. The use of the land within the garden

The cultivation of examples of native vegetation, principally tree species of the Atlantic Forest, which are little known or those which are rare or threatened, has priority over the the cultivation of other plants. Also planted are species from other Brazilian ecosystems, exotic species of economic importance, ornamental or scientific importance. Given that there are as yet no greenhouses or other plant houses, priority is given to the cultivation of tree species. As the number of employees needed is insufficient, a large part of the work is done by university students who register as trainees and do eight hours of work per week.

Working and maintaining the physical area is extremely difficult given the acidity of the soil and the large quantity of weeds. In the tropics, and particularly in the months of summer and autumn, weeds grow very rapidly and strangle cultivated species. In the months of December, January and February all the work in the botanic garden is focussed on pruning, on the control of invasive plants, of ants and of predatory lizards. This work, without the use of chemicals stimulated the study of the biological control of pests and is especially valuable for raising the awareness of students to the necessity of the serious study of this subject.

Criticism of the use of agrotoxic chemicals is very common and all too easy. To control pests in tropical regions is rare and complex. In this regard as the botanic garden is a university one giving priority to agricultural matters, it must be in the forefront in promoting natural pest control.

2. Development of an orientation programme for visiting students and professors.

A botanic garden which is part of a university has a very important educational function. We do not have any courses on the conservation of fauna and flora, or on environmental law. The botanic garden therefore becomes responsible for the conservation education of the various professionals trained in the university.

Professionals working in public and private institutions who studied at the university often come back to their original university, with the aim of consulting their former teachers, to gain new knowledge and to used the equipment and material in the university. Through these people, the botanic garden maintains a link with the many varied communities outside of the university. Through these ex-students, the botanic garden can influence the public and government in many ways.

Similarly, the botanic garden is of importance, because it plays a part in the training of a large number of professionals and can act as a source of information and enlightenment for professional scientists in problems relating to the conservation of flora and environmental balance.

3. The collection and cultivation of rare native tree species, especially those in danger of extinction.

One of the uses to which the land area of the botanic garden is put is the very necessary collection and cultivation of seeds. As well as that, the botanic garden maintains a nursery of plants raised from these seeds which are either planted in the botanic garden and the university campus or else sold to the general public. The public has therefore the opportunity to buy unusual species which they can use for landscaping and arboriculture. The botanic garden hopes that this activity may help in a few years to ensure that landscaping becomes a feature of many park and gardens. It is hoped that it will also have an effect by increasing the numbers of native plants that are grown, and stimulate reforestation effort using native, not exotic plants. The work of the nursery is done with the collaboration of students. It is interesting to note that the general public has created a demand for native tree species for planting. Notices are circulated by us on forest fires, rare species, medicinal plant usage and flora devastation, and these have awakened the interest in people about endangered plants. Commercial nurseries, on the other hand, frequently concentrate on the cultivation of traditionally better known species, which are easily marketed. Thus the botanic garden with this secondary, but nevertheless valuable activity, is able to supply the public and the mayors of various municipalities of the state of Rio de Janeiro with new options for tree planting and reforestation. The activity is becoming very important. It is also valuable that the students in caring for and handling native species learn to recognize them and by the time they leave university will have acquired useful knowledge about the native flora.

Principal difficulties

1. The lack of funds for field trips means that the collection of seeds is unreliable. It is carried out mainly by botany, zoology and ecology teachers when they go on field trips. It is also carried out by students during the holidays in various Brazilian localities. This random means of collection makes it impossible to have a stable plan of activities for the nursery and to have a permanent catalogue of cultivated species.

2. The lack of greenhouses and of an adequate irrigation system makes it impossible to cultivate herbaceous plants. Especially difficult is the culture of plants of families such as the Piperaceae, Gesneriaceae and Orchidaceae, as well as other families that are common in the Atlantic Forest. Many of these families contain numerous species that are on the brink of extinction.

3. The lack of an information system and database makes it difficult for us to communicate with other gardens or with the general public.

Resumo

A Universidade Federal Rural do Rio de Janeiro, sediada no municîpio de Itaguaí, a 60 km do Rio de Janeiro, mantem cursos de graduaçao em Agronomia, Engenharia Florestal, Veterinaria, Zootecnia, Ciéncias Agrícolas, Economia Doméstica, Biologia, Química, Física e Matemática. Mantém ainda um Colégio Técnico Agrícola que forma profissionais de nivel médio e cursos de pós-graduaçao em parasitologia Veterinaria, Microbiologia, Fisiologia de Solos e Química. O 'campus' com edificaçoes amplas, em estilo europeu, tem várias áreas experimentais pertencentes aos diferentes setores de ensino.

Há oito anos atraz, por sugestao e pressao dos professores do Instituto de Biologia, especialmente do Departamento de Botânica foi demarcada uma área de 18 hectares na qual iniciou-se a implantaçao de um Jardim Botânico. Ná área demarcadea, havia uma residencia que foi adaptada ao funcionamento da sede do Jardim Botânico. Iniciou-se entao a construçao de cercas em torno da área, a abertura de estradas, a construçao de um lago e o planejamento geral de plantio. Oito funcionários foram cedidos por diversos setores da Universidade e compoem hoje o quadro funcional do Jardim Botânico que está ligado administrativamente a reitoria, o que lhe confere bastante autonomia de funcionamento. A orientaçao e o trabalho tecnico é feito por professores do Departamento de Botânica, que dirigem, fazem o trabalho de curadoria do herbário e o planejamento geral do orgao.

Nos primeiros anos de existencia o Jardim Botânico centrou suas atividades em trés campos: 1) ocupaçao da área física; 2) Atendimento e orientaçao de estudantes e professores; 3) Coleta e cultivo de espécies arbóreas nativas raras e/ou ameaçadas de extinçadas de extinçao.

A formaçao e o treinamento de alunos em atividades de coleta e plantio de espécis nativas e no conhecimento e conservaçao da flora, especialmente do ecossisstema Floresta Atlantica tem sido uma das atividades prioritárias. Por ter acesso a um grande numero de profissionais em formaçao, o Jardim Botânico como parte de uma Universidade, torna-se importante como fonte de informaçao de profissionais diversos nos segredos da manutençao de espécies nativas, da conservaçao da flora e do equilibrio ambiental. Como parte de uma Universidade predominantemente agricola tem um importante papel na formaçao de profissinais que atuarao nos orgaos e empresas ligadas a agricultura e meio ambiente.

Rôle of botanic gardens in Sri Lanka and the conservation of medicinal plants

D.B. Sumithraarachchi

Royal Botanic Gardens
Peradeniya, Sri Lanka

Introduction

Sri Lanka is a continental island situated in the Indian Ocean near the southern tip of mainland India. The island is small around 6.5 million ha. in size with high plant diversity. The diverse nature of the vegetation have been recognised by the state and a large number of national parks and protected areas have been declared and protected since 1938. Protected land areas up to 1986 have been around 13% of the country's land surface. In addition to the well known rainforests, these areas include some of the important grasslands and scrub jungles. Plant diversity in Sri Lanka, in terms of endemism, has been recorded as 24% for flowering plants, and 18% for pteridophytes. Although there are no records of endemic angiosperm families for Sri Lanka, 25 genera have been recorded as endemic to the island.

The various plant groups found in Sri Lanka can be summarised as follows:-

	Genera	Species	Endemics
Mosses	168	519	?
Liverworts	64	181	?
Pteridophytes	97	314	57
Embryophytes	1290	3300	815

The Botanic Gardens in Sri Lanka are situated in 3 ecological zones and are visited by a large number of people.

TROPICAL BOTANIC GARDENS: THEIR ROLE IN CONSERVATION AND DEVELOPMENT
ISBN 0—12—346850—7

		Royal Botanic Gardens Peradeniya 1821	Botanic Gardens Hakgala 1861	Botanic Gardens Gampaha 1876
1.	Ecological zone	Mid country Intermediate to wet zone	Hill country Montane zone	Low country Wet zone
2.	Area	62 ha.	24 ha	15 ha
3.	Altitude	500 m.	2,000 m.	around sea level
4.	Plant collection	4,800 species	1,000 species	500 species
5.	Number of endemic plants in collection	99 species	60 species	38 species
6.	*Ex situ* collections	81 species of orchids + over 100 other species	20 species of orchids + over 25 other species	10 species of orchids + over 25 others
7.	Medicinal plants	Over 300 species	Over 40 species	Estimated to be more than 60
8.	Visitors	638,219	132,934	96,395
9.	Scientific staff	8	1	2

Botanic gardens

The Treasury in Sri Lanka has allocated the following functions to botanic gardens.

1. Maintenance and development of botanic gardens.
2. Maintenance and development of national herbaria
3. Research and development of floriculture.
4. Education and extension of ornamental horticulture, floriculture, plant conservation and environmental improvements.

5. Care and maintenance of the Commonwealth cemeteries in Sri Lanka.
6. Care and maintenance of garden attached to the official residences and offices of the H.E., The President and Hon. Prime Minister.

The main objectives of the botanic gardens have been revised using the recommendations made at the Las Palmas Congress in 1985, and subsequently the Botanic Gardens Conservation Strategy (WWF/IUCN-BGCS). The botanic gardens have been fortunate in attracting 12 university students to help the scientific activities recently, due to the temporary closure of the universities. The students are assisting in field surveys, especially of two important forest areas known as the Knuckles range of mountains and the Peak Wilderness. The Knuckles range cover about 18,400 ha. and the Peak Wilderness cover 22,400 ha. These forests are protected by law. But there are many practical problems and the IUCN has given assistance to study a management pattern for the Knuckles range following visits made by Prof. V.H. Heywood. The botanical surveys in progress reveal that these forests are extremely interesting and house a large number of plants endemic to the specific forests with another considerable number being shared between these two forests and the well known Sinharaja Forest. The discoveries of these surveys, when completed, will be the subject of a future congress. These activities indicate to us the diversity of plants in what is known as the mid-elevational vegetation of Sri Lanka. Most of the mid-elevational forests have been lost through conversion into tea and rubber plantations since the 19th century.

The survey of nitrogen-fixing plans in the ecosystem is another important activity that is carried out by the botanic gardens. The results of this survey is expected to highlight plant species for the improvement of degraded environments.

The collection of rare and endangered palms, orchids and other plant species is another activity area in which the limited staff is progressing successfully.

The Biotechnology Laboratory produced many species of orchids for orchid enthusiasts and for re-introduction into natural forests.

Special palms such as *Loxococcus rupicola, Areca concinna, Lodoicea maldivica* are now available for limited circulation to botanical institutions.

In addition to these activities the botanic gardens in Sri Lanka carry out educational programmes for school children, school leavers and employees on the importance of environmental protection and aspects related to plant conservation.

Importance of medicinal plants

The main subject proper of this paper is the conservation of medicinal plants. Sri Lanka is inhabited by 16.1 million people and it is estimated that more than 20% of the population is dependent upon the traditional herbal system of medicine.

Present natural forest tree cover in Sri Lanka is estimated to be around 9%, although the forest cover including plantation forests adds up to an attractive 29%. The protected area cover in the whole country is estimated at 13.38% which amounts to some 878,986 ha. The Kandy district, in which the Royal Botanic Garden, Peradeniya is located, is 193,950 ha. in size with a population of around 1.1 m.

In Sri Lanka there are about 9,000 registered general practitioners of herbal medicine with about another 4,000 specialists. In the Kandy district alone there were 460 registered herbal practitioners with about 200 practising at present. The others in the list have been discovered as persons not practising at present due to old age, death, change of residence etc. Over 200,000 people in the Kandy district are estimated to be using plants as a source of medicine. On many occasions the western doctors recommend the use of certain herbal treatments through specialist herbal doctors. It is important to appreciate that the herbal doctors are recognised as valuable by the western medical doctors.

In the Kandy district alone it is also estimated that at least 200 metric tons of dried medicinal plants and plant products are needed every year to treat these patients. Most of these plants will come from natural habitats, home gardens and disturbed areas in Sri Lanka. They may not always come from within the district. For example *Coscinium fenestratum* used for Berberin would come from outside the Kandy district and is a heavily used plant and one in danger of overexploitation. It is also important to note that the extraction of roots of plants in the past have been by uprooting the plant.

A proposal was made to conserve medicinal plants in Sri Lanka and this was accepted by WWF for funding. The objective of the study proposal was:

1. The preparation of check lists of medicinal plants, after proper identification.
2. Preparation of legislative enactments to protect the medicinal plants in their natural habitats against overexploitation.
3. Identification and declaration of sites for *in situ* conservation. An island-wide survey was designed, on a district basis, with some special historic localities to be studied in detail.

The activities reported here refer to the survey carried out in the district of Kandy.

DISTRICT BOUNDARIES OF SRI LANKA

The survey was designed along the following lines.

1. Analysis of plant extraction patterns from the environment
2. Preparation of an inventory of plants available in the Kandy district
3. Enumeration of the constraints for the conservation of medicinal plants
4. Recommendations.

Plant extraction patterns

The survey was started with a fact-finding questionnaire in the area of survey. Annex 1 indicates the results of the first phase of this survey. The survey was carried out by sending the questionnaire to the herbal doctors by post and subsequently visiting them and having personal interviews.

Herbs for treatments were found to be recommended by the practitioners in different forms.

a. Freshly collected herbs from forests, home gardens etc.
b. Dried plants and parts that can be purchased from a market shop.
c. Preparation of herbal extractions from factories.

The survey brought up some astonishing discoveries: a lot of apparently miraculous treatments on the one hand and total failures on the other. The apparent miracles included fracture healing, treatments of asthma, hayfever and allergies. The disappointments have been taken by us (botanists) very seriously: The use of the wrong plants, lack of help and knowledge to recognise the plants and the lack of standards, especially in the use of poisonous plants such as *Strychnos nux-vomica* etc. This was common and encountered at both practitioner and patient level.

Plant inventories

The survey enumerated 280 species growing in the Kandy district used in herbal medicine (Annex 2). The survey listed all the botanically recognised plants used in herbal medicine which amounted to 562.

130 out of 282 species were found to be heavily used and 64 plant species were in danger of extinction from their natural habitats due to continuous extraction, as indicated in Annex 2 column 2.

The survey revealed that 26 species of plants seen in Annex 2 are present day weeds and are heavily used in herbal treatments. The data collected are available in the Royal Botanic Gardens economic plants database.

Annex 3 contains a list of medicinal plants cultivated at the Royal Botanic Gardens, Peradeniya.

Constraints

1. The biggest constraint was the lack of knowledge in recognising the actual plants. This, surprisingly, was encountered with the newer generations and also led to overexploitation of valuable plants from natural habitats.
2. The herbal doctors could not get fresh plants and parts of plant for their treatments due to the depletion of plants from their neighbourhood. To this was attributed the decrease in the quality of treatments.
3. The practitioners themselves did not have the land, time and enthusiasm to cultivate medicinal plants in their home gardens, although they indicated their desire to do so if the state provided them with land.
4. The dried plants and parts of plants available for sale in the market were often found to be adulterated.
5. Loss of collected plants due to improper storage.

The cultivation of medicinal plants was limited due to a number of reasons:

a. Lack of a commercial market.

b. Lack of knowledge in methods of propagation, cultivation and preservation.

c. Lack of land for cultivation.

During the survey it was found that the actual extraction of plants from the natural forests by the knowledgeable practitioners was limited.

Achievements and recommendations

1. Cultivation of medicinal plants with state support. Along with the survey of plant use pattern, action was taken to study the methods of propagation of medicinal plants in commercial quantities. The achievements of the botanic gardens in this area was commended by the Hon. Minister of Indigenous Medicine.

   The botanic gardens directed the establishment of a 15 ha. garden for the sole purpose of propagating and cultivating medicinal plants. This was established in 1986. This garden is now producing over 50 species of medicinal plants that are in high demand and are available to those interested in the whole island. In collaboration with the botanic gardens experiments were also carried out on traditional systems of propagation, and embryo culture where traditional systems of propagation took long periods of time.

2. The late Superintendent of botanic gardens, Mr D.M.A. Javaweera has produced a 5-volume series on medicinal plants of Ceylon. In addition to this valuable taxonomic compilation, a small handbook has been produced for herbal practitioners.

3. A booklet on the propagation and cultivation of medicinal plants has also been compiled and is awaiting printing and publication.

4. A database has been developed for medicinal plants of the Kandy district. This should be extended to the whole island.

5. The protection of hitherto unprotected natural forest patches used by herbal doctors in the Kandy district. These must be protected as sites for medicinal plant diversity.

Acknowledgements

I am grateful to the following persons for making this task possible. Mr Lyn de Alwis - Sri Lankan Coordinator of WWF 3320 Project. Mr L. Sugunadasa, Commissioner of Ayurveda. Ms K. Tennakoon, M. Rizvi, Ms D. Tilakaratne, Ms U. Wickramasinghe, Research Assistants. The Ayurvedic Practitioners who supported our surveys. Ms S. Amanulla, computer assistant for all the computer work. Mr Hugh Synge and Ms

Pascale Moehrle of WWF for all the encouragement and financial support to allow me to attend the Second International Botanic Gardens Conservation Congress in Réunion. I also wish to extend my sincere thanks to all my colleagues who helped me in many ways.

References

Abeywickrama, B.A. The genera of Ceylon pteridophytes. *Ceylon Jour. Sci.* Vol. XIII. pt 1.

Abeywickrama, B.A. (1959). The genera of the liverworts of Ceylon. *Ceylon Jour. Sci.* Vol.2.1. p.33-81.

Abeywickrama, B.A. (1960). The genera of mosses of Ceylon, *Ceylon Jour. Sci. Biol. Sci.* Vol. 3.1.: p.41-120.

Trimon, H. (1890 - 1986). *Handbook to the Flora of Ceylon* Vol. I-V.

Dassanayake, M.D. & Fosberg, F.R. *Revised Handbook to the Flora of Ceylon* Vol. 1 - 6.

Jayaweera, D.M.A. (1981). *Medicinal Plants used in Ceylon*, Parts 1-5.

Gunatilleke, C.V.S., Gunatilleke, I.A.U.N. and Sumithraarachchi, D.B. (1987). Woody endemic species of wet low lands of Sri Lanka and their conservation in botanic gardens. In: *Botanic Gardens and the World Conservation Strategy*. Ed. Bramwell D., Hamann O., Heywood V. and Synge H., p. 183-194. Academic Press, London and New York.

Sumithraarachchi, D.B. (1985). *Training Manual for Traditional Practitioners*. WHO Report.

Sumithraarachchi, D.B. (1986). *Development of Medicinal Plant Nursery*. WHO Report.

Annex 1
Medicinal Plant Extraction Pattern - Kandy District

[Prepared according to the answers given by 163 Ayurvedic Physicians to our questionnaire]

1. The ordinary medicinal plants are obtained from

1. Surroundings	79.75%
2. Other areas	34.36%
3. Forests	36.51%
4. Market	67.48%

2. Special medicinal plants are collected by

1. The physician himself	34.36%
2. The physician and his family members	50.31%
3. Friends and relatives	32.52%
4. Outsiders	67.48%

3. The difficulties met with obtaining medicinal plants.

1. The difficulty of identification	23.93%
2. Due to extinction within the area	44.17%
3. Due to extinction in the country	31.90%
4. Due to unscrupulous collection	34.97%
5. Due to forest destruction	55.83%
6. Due to any other reason (e.g.agriculture)	11.66%

4.1 Although the medicinal value of naturally grown plants can be obtained by cultivating them, the plants are not cultivated because of

1. Lack of suitable land	71.17%
2. Lack of technical knowhow in propagation	25.77%
3. Lack of technical knowhow in cultivation	23.93%
4. Lack of time for cultivation	14.11%
5. No need for cultivation	4.29%
6. The necessary plant can be obtained from other areas	9.20%

4.2 The medicinal value of naturally grown plants cannot be obtained by cultivating them. These plants are therefore obtained from:

1. Natural forests and reserves	30.06%
2. The market	34.97%

Annex 2.
Utilization of medicinal plants in the Kandy District

SPECIES	Heavily used species	Species extracted from natural forests	Present day weeds
Abrus precatorius	x	x	
Acalypha indica	x		x
Acorus calamus	x		
Acronychia pendunculata	x	x	
Achyranthes aspera	x		x
Adenanthera pavonia	x		
Adhatoda vasica	x		
Adiantum caudatum			
Adina cordifolia			
Aegle marmelos	x		
Aerva lanata	x		x
Aglaia roxburghiana			
Allium sativum	x		
Allophylus cobbe			
Alocasia indica			
Aloe vera	x	x	
Alstonia scholaris	x	x	
Alternanthera sessilis	x	x	x
Alysicarpus vaginalis	x		
Amaranthus spinosus			
Amaranthus viridis			
Amoora rohituka			
Amorphophallus campanulatus	x		
Ampelocissus indica			
Anacardium occidentale			
Anamirta cocculus			
Andrographis paniculata			
Annona squamosa			
Anoectochilus setaceus	x	x	
Antiaris toxicaria			
Aponogeton crispus	x		
Aporosa cardiosperma	x		
Areca catechu			
Argemone mexicana			x
Arisaema leschenaultii			
Aristolochia indica	x	x	
Artemisia vulgaris			
Artocarpus heterophyllus			

SPECIES	Heavily used species	Species extracted from natural forests	Present day weeds
Asparagus falcatus	x		
Averrhoa bilimbi			
Azadirachta indica	x		
Bacopa monnieri	x		
Baliospermum montanum	x		
Bambusa arundinacea			
Barleria prionitis	x		x
Bauhinia acuminata			
Bauhinia tomentosa	x	x	
Biophytum reinwardtii			
Blumea millis			
Brassica integrifolia			
Bridelia retusa			
Brucea javanica			
Caesalpinia digyna			
Caesalpinia bonduc	x	x	
Callicarpa tomentosa			
Calotropis gigantea	x		x
Canna indica		x	x
Cannabis sativa	x		
Canscora decussata			
Capparis horrida	x	x	
Capparis moonii			
Cardiospermum microcarpum	x		
Casearia zeylanica			
Cassia alata	x	x	
Cassia auriculata	x		
Cassia fistula	x		
Casuarina equisetifolia	x		
Ceiba pentandra			
Celtis cinnamomea			
Centella asiatica	x	x	
Centranthera indica			
Cinnamomum zeylanicum	x		
Cissampelos pareira	x	x	
Cissus quadrangularis	x	x	
Citrus aurantifolia	x		
Clerodendrum infortunatum			
Clerodendrum serratum			
Clitoria ternatea	x		
Coccinea grandis			

SPECIES	Heavily used species	Species extracted from natural forests	Present day weeds
Cocos nucifera			
Coix lacryma-jobi			
Coleus amboinicus	x		
Colocasia esculenta			
Commelina benghalensis		x	x
Connarus monocarpus	x		
Cordia dichotoma			
Coscinium fenestratrum	x	x	
Costus speciosus			
Crateva religiosa	x		
Croton aromaticus			
Croton lacciferus			
Cryptolepis buchananii			
Cucumis callosus	x	x	
Curcuma aromatica			
Curcuma domestica	x		
Cuscuta chinensis			
Cyperus rotundus	x		x
Datura metel	x		x
Desmodium heterophyllum			
Desmodium triflorum	x		x
Dregea volubilis	x	x	
Drosera burmanii			
Drosera indica	x		
Drymoglossum heterophyllum			
Eclipta prostrata	x		x
Elephantopus scaber	x		x
Eleusine indica		x	x
Emilia sonchifolia			
Ephemerantha macraei			
Ervatamia divaricata			
Eryngium foetidum			
Erythroxylum monogynum	x		
Erythroxylum moonii			
Eugenia caryophyllata			
Eupatorium triplinerve			
Euphorbia hirta			
Euphorbia neriifolia			
Euphorbia thymifolia			
Euphorbia tirucalli	x		
Evolvulus alsinoides			

SPECIES	Heavily used species	Species extracted from natural forests	Present day weeds
Feronia limonia	x		
Ficus benghalensis			
Ficus hispida			
Ficus racemosa	x	x	
Ficus religiosa			
Flacourtia ramontchi			
Fleurya interrupta	x	x	x
Garcinia mangostana			
Garcinia morella			
Geophila herbacea			
Girardinia zeylanica			
Glycosmis pentaphylla			
Gmelina arborea	x		
Grewia tiliifolia	x	x	
Gymnema sylvestre	x	x	x
Hedyotis auricularia			
Hedyotis fruticosa			
Heliotropium indicum			
Hemidesmus indicus	x	x	x
Heteropogon contortus	x	x	x
Hibiscus abelmoschus	x	x	
Hibiscus rosa-sinensis			
Holarrhena antidysenterica	x	x	
Holarrhena mitis	x		
Horsfieldia irya			
Hydnocarpus venenata			
Hydrocotyle javanica			
Hypserpa nitida			
Imperata cylindrica			
Indigofera tinctoria	x		
Ipomoea angustifolia			
Ipomoea asarifolia	x		
Ipomoea mauritiana	x		
Ipomoea nil			
Ipomoea obscura			
Ipomoea pes-tigridis	x	x	x
Ixora coccinea	x	x	
Jasminum angustifolium			
Jatropha podagrica			
Justicia betonica			
Justicia procumbens			

SPECIES	Heavily used species	Species extracted from natural forests	Present day weeds
Kaempferia galanga	x		
Kalanchoe laciniata			
Kokoona zeylanica			
Lagerstroemia speciosa			
Languas calcarata	x		
Languas galanga	x		
Lasia spinosa	x		
Leucas marrubioides			
Leucas zeylanica			
Litsea longifolia			
Litsea glutinosa	x	x	
Luffa acutangula			
Luffa cylindrica			
Lycopodium cernum	x	x	
Lycopodium phlegmaria	x	x	
Madhuca longifolia	x		
Madhuca neriifolia	x		
Mangifera indica			
Marsdenia tenacissima	x	x	
Melastoma malabaricum			
Memecylon umbellatum			
Mentha arvensis			
Mentha sylvestris			
Michelia champaca			
Monochoria hastata		x	x
Morinda citrifolia	x	x	
Morinda tinctoria	x	x	
Moringa oleifera	x		
Munronia pumila	x	x	
Mussaenda frondosa			
Myristica fragrans			
Naravelia zeylanica			
Nardostachys jatamansi	x	x	
Nauclea orientalis	x	x	
Neolitsea involucrata			
Nymphaea stellata	x	x	
Ocimum americanum			
Ocimum gratissimum			
Ocimum sanctum			
Operculina turpethum	x	x	
Ophiorrhiza mungos	x	x	

SPECIES	Heavily used species	Species extracted from natural forests	Present day weeds
Oroxylum indicum	x	x	
Osbeckia aspera			
Osbeckia octandra			
Oxalis corniculata			
Paramignya monophylla			
Paspalum conjugatum			
Paspalum scrobiculatum			
Pavetta indica	x		
Pergularia daemia			
Phaseolus adenanthus			
Phyllanthus debilis	x		x
Phyllanthus emblica	x	x	
Piper betle			
Piper chuvya	x		
Piper longum	x		
Piper nigrum	x		
Pisonia grandis			
Plectranthus zeylanicus	x	x	
Plumbago indica	x	x	
Plumbago zeylanica	x	x	
Pogostemon heyneanus	x		
Polygonum barbatum			
Polygonum chinense			
Pongamia pinnata	x	x	
Portulaca quadrifida			
Pothos scandens	x	x	
Premna serratifolia	x		
Premna tomentosa			
Psidium guajava			
Pterocarpus marsupium	x	x	
Pterocarpus santalinus	x		
Pterospermum canescens			
Punica granatum			
Rauvolfia serpentina	x	x	
Ricinus communis	x		
Rungia repens	x		
Salacia reticulata	x	x	
Salmalia malabarica			
Santalum album	x	x	
Saraca indica	x	x	
Semecarpus gardneri			

SPECIES	Heavily used species	Species extracted from natural forests	Present day weeds
Semecarpus obscura			
Sesbania grandiflora			
Sida racemosa	x		x
Sida rhombifolia			
Sida veronicaefolia	x		x
Smilax zeylanica			
Solanum ferox			
Solanum indicum			
Solanum surattense	x		x
Solanum trilobatum	x		
Solanum xanthocarpum	x		
Solanum nigrum			
Spilanthes paniculata			
Stephania japonica			
Stereospermum suaveolens	x		
Streblus asper			
Strychnos potatorum	x	x	
Symplocos loha			
Syzygium cumini	x	x	
Tarenna asiatica	x	x	
Tephrosia purpurea			
Terminalia arjuna	x	x	
Terminalia bellirica	x	x	
Terminalia chebula	x	x	
Toddalia asiatica			
Trianthema decandra	x		
Tribulus terrestris	x		x
Tylophora indica			
Ventilago madraspatana			
Vetiveria zizanioides	x	x	
Vitex negundo	x	x	
Withania somnifera	x		
Woodfordia fruticosa			
Zanthoxylum tetraspermum			
Zeuxine regia	x		
Zingiber zerumbet	x		
Ziziphus rugosa			

Annex 3. Medicinal plants cultivated in the Royal Botanic Gardens Peradeniya

Acanthaceae
Blepharis repens
Barleria lupilina
Justicia procumbens
Rungia repens
Andrographis paniculata
Adhatoda sp.
Barleria prionitis
Adhatoda vasica

Aizoaceae
Mollugo pentaphylla

Alangiaceae
Alangium salvifolium

Amaranthaceae
Amaranthus tricolor
Amaranthus paniculatus
Achyranthes sp.
Amaranthus spinosus
Celosia argentea
Achyranthes aspera
Cyathula prostrata
Aerva lanata
Amaranthus viridis

Amaryllidaceae
Pancratium zeylanicum
Crinum zeylanicum
Crinum asiaticum
Allium sativum
Allium cepa
Allium ascalonicum

Apocynaceae
Carissa carandas
Strophanthus sarmentosus
Ervatamia divaricata
Rauvolfia serpentina
Strophanthus preussi
Nerium oleander

Araceae
Arisaema leschenaultii
Amorphophallus campanulatus
Alocasia sp.
Pothos scandens
Acorus calamus

Aristolochiaceae
Aristolochia indica

Asclepiadaceae
Hoya ovalifolia
Gomphocarpus fruticosus
Wattakaka volubilis
Marsdenia tenacissima
Hoya carnosa
Gymnema sylvestre
Calotropis gigantea
Hoya rubida
Holostemma annulare

Balsaminaceae
Impatiens repens

Basellaceae
Basella alba

Boraginaceae
Trichodesma zeylanicum
Heliotropium indicum
Carmona microphylla

Cactaceae
Opuntia dellenii
Epiphyllum crenatum

Capparidaceae
Gynandropsis gynandra
Polanisia icosandra
Cleome viscosa

Caricaceae
Carica papaya

Chenopodiaceae
Chenopodium ambrosiodes

Commelinaceae
Cyanotis zeylanica
Commelina benghalensis
Cyanotis fasciculata

Compositae
Melampodium divaricatum
Emilia sonchifolia
Centipeda minima
Spilanthes paniculata
Ageratum conyzoides
Anaphalis subdecurrens
Anaphalis zeylanica
Eupatorium ayapana
Elephantopus scaber
Vernonia cinerea
Vernonia anthelmintica
Wedelia chinensis
Pyrethrum cineraefolium
Eclipta prostata
Acanthospermum hispidum
Artemisia dubia

Convolvulaceae
Merremia tridentata
Ipomoea mauritiana
Ipomoea triloba
Ipomoea nil
Evolvulus alsinoides
Merremia cissoides
Operculina turpethum
Ipomoea asarifolia
Ipomoea sepiaria
Ipomoea pes-caprae

Crassulaceae
Kalanchoe pinnata
Sedum sexangulare
Kalanchoe laciniata

Cruciferae
Brassica integrifolia

Cucurbitaceae
Momordica dioica
Zanonia indica
Momordica charantia
Cucumis callosus
Coccinia grandis

Cyperaceae
Cyperus corymbosus
Cyperus rotundus

Erythroxylaceae
Erythroxylum moonii

Euphorbiaceae
Croton tiglium
Jatropha curcas
Jatropha podagrica
Tragia involucrata
Jatropha integerrima
Acalypha indica
Phyllanthus pulcher
Pedilanthus tithymaloides
Euphorbia antiquorum
Phyllanthus debilis
Euphorbia neriifolia
Securinega leucopyrus
Jatropha multifida
Ricinus communis
Phyllanthus reticulatus
Croton aromaticus
Sauropus androgynus
Euphorbia indica
Jatropha gossypifolia
Croton lacciferus
Euphorbia tirucalli

Gentianaceae
Exacum trinerve

Gramineae
Coix lacryma-jobi
Cymbopogon citratus
Panicum antidotale
Saccharum officinarum
Vetiveria zizanioides
Cymbopogon nardus
Cymbopogon confertiflorus
Cymbopogon winterahus

Labiatae
Orthosiphon aristatus
Leonotis nepetaefolia
Plectranthus zeylanicus
Coleus malabaricus
Coleus ambfinicus
Leucas zeylanica
Pogostemon heyneanus
Coleus rotundifolius
Leucas marrubioides
Pogostemon parviflorus
Anisomeles indica
Mentha sylvestris

Lamiaceae
Ocimum sanctum

Leeaceae
Leea indica

Leguminosae
Wagatia spicata
Cassia surattensis
Cassia occidentalis
Cajanus cajan
Desmodium heterophyllum
Alysicarpus vaginalis
Cassia fistula
Caesalpinia bonduc
Desmodium gyrans
Cassia auriculata
Indigofera tinctoria
Phaseolus radiatus
Tephrosia purpurea
Clitoria ternatea
Desmodium triflorum
Abrus precatorius
Desmodium heterocarpum
Cassia alata
Crotalaria laburnifolia

Mucuna prurita
Phaseolus mungo
Dolichos biflorus
Cassia tora

Liliaceae
Aloe vera
Sansevieria zeylanica
Gloriosa superba
Asparagus racemosus
Aloe vera var. *littoralis*

Lycopodiaceae
Lycopodium phlegmaria

Lythraceae
Woodfordia fruticosa

Malvaceae
Hibiscus abelmoschus
Sida veronicaefolia
Sida cordifolia
Hibiscus sabdariffa
Gossypium herbaceum
Hibiscus micranthus
Wissadula periphocifolia
Urena obata
Hibiscus furcatus
Sida accuta
Abutilon indicum

Marantaceae
Maranta arundinacea

Melastomataceae
Osbeckia aspera
Melastoma malabathricum
Memecylon umbellatum

Meliaceae
Munronia pumila
Azadirachta indica

Menispermaceae
Tinospora cordifolia
Cyclea burmanni
Tinospora malabarica
Tinospora glabra
Anamirta cocculus
Cissampelos pareira
Stephania japonica

Moringaceae
Moringa oleifera
Embelia ribes

Nyctaginaceae
Mirabilis jalapa
Pisonia grandis
Boerhavia diffusa

Oleaceae
Nyctanthes arbor-tristis
Jasminum officinale var. *grandiflorum*
Jasminum grandiflorum
Jasminus pubescens
Jasminum revolutum
Jasminum angustifolium
Jasminum sambac

Ophioglossaceae
Helminthostachys zeylanica

Oxalidaceae
Oxalis corniculata
Biophytum nervifolium

Papaveraceae
Argemone mexicana
Adenia palmata

Pedaliaceae
Sesamum indicum

Piperaceae
Piper nigrum
Piper longum
Piper chuvya
Piper siriboa
Piper umbellatum
Piper betle

Plumbaginaceae
Plumbago zeylanica
Plumbago indica
Polygonum chinense

Polypodiaceae
Drynaria quercifolia
Drymoglossum heterophyllum
Adiantum caudatum

Portulacaceae
Portulaca oleracea
Portulaca quadrifida

Punicaceae
Punica granatum

Ranunculaceae
Clematis gouriana
Clematis flammula
Naravelia zeylanica

Rubiaceae
Tarenna asiatica
Borreria hispida
Oldenlandia corymbosa
Hedyotis auricularia
Morinda tinctoria
Ophiorhiza mungos
Pavetta indica
Gardenia latifolia
Mussaenda frondosa
Paederia foetida
Ixora coccinea

Rutaceae
Acronychia pedunculata
Murraya koenigii
Paramignya monophylla
Ruta graveolens
Clausena indica
Glycosmis pentaphylla
Triphasia trifolia
Toddalia asiatica

Santalaceae
Santalum album

Sapindaceae
Cardiospermum halicacabum
Allophylus cobbe

Scrophulariaceae
Bacopa monniera
Artanema longifolium

Solanaceae
Solanum trilobatum
Solanum nigrum
Solanum surattense

Datura suaveolens
Solanum indicum
Datura metel

Tiliaceae
Clappertonia ficifolia

Umbelliferae
Centella asiatica
Eryngium foetidum
Hydrocotyle sibthropioides
Coriandrum sativum
Cuminum cyminum
Apium graveolens
Foeniculum vulgare

Verbenaceae
Clerodendrum infortunatum
Phyla nodiflora
Clerodendrum serratum
Premna obtusifolia
Vitex negundo
Callicarpa tomentosa
Vitex trifolia

Vitaceae
Cissus vitiginea
Cayratia pedata
Cissus quadrangularis
Zingiberaceae
Elettaria cardamomum var. *cardamomum*
Languas vkttata
Boesenbergia rotunda
Hedychium coccineum
Kaempferia galanga
Alpinia galanga
Hedychium coronarum
Zingiber cylindricum
Kaempferia rotunda
Costus speciosus
Curcuma longa
Zingiber officinale
Curcuma zedoaria

Cultivation on the French Mediterranean coast (the Côte d'Azur) of rare or threatened tropical or sub-tropical trees, two examples: *Jubaea chilensis* and *Cupressus dupreziana*

Pierre Auge

l'Association des Responsables et Techniciens de Jardins Botaniques, France

Catherine Ducatillion

Jardin Botanique de la Villa Thuret, Institut National de la Recherche Agronomique, France

Summary

The rôle of the Botanic Garden "Villa Thuret" (Antibes, France) is to introduce and establish subtropical woody plants. In addition, one of the aims of this garden is to exhibit species that are uncommon or threatened in their natural habitat. For example, the fruiting of about ten individuals of *Jubaea chilensis* allows us to collect and distribute several thousands of seeds each year. Moreover, development of collaboration with other institutions has allowed us to collect 285 *Cupressus dupreziana* individuals. These trees, arising from vegetative or sexual reproduction, are planted into a forest arboretum where they can be studied *ex situ* and used as a source of seeds.

Introduction

For more than a century the Côte d'Azur has been a privileged place for the introduction and establishment of exotic plant species. Its climate,

TROPICAL BOTANIC GARDENS: THEIR ROLE IN CONVERVATION AND DEVELOPMENT
ISBN 0—12—346850—7

the shape of its shoreline, the diversity of its soil, has moulded a mosaic of micro-environments, favourable to the establishment of foreign plants. These introductions have contributed to the creation of the present day plant landscape, and has given it great originality. Many gardens, which in the past assembled rich collections of plants, have today disappeared; one of the oldest, the Jardin de la Villa Thuret (named after its creator) still exists today, having resisted the erosion which caused other gardens to disappear.

The garden of the Villa Thuret was bequeathed to the nation in 1875, so that botanical research could be undertaken there. Today it is managed by the Institut National de la Recherche Agronomique. Its primary rôle to this day has been in conservation, the introduction and the study of woody plants or legumes in the open air, and in particular for the diversification of horticulture and sylviculture.

Among the plants introduced, Thuret and his successors, collected some rare species and this work has continued to the present day. The garden contains some species which feature on the list of rare or threatened plant species as prepared by IUCN.

For example, in the palm family we can mention the following examples:

Brahea edulis H.A. Wandland (Syn. *Erythea edulis* S. Watson), from the island of Guadeloupe
Chamaedorea microspadix Burnet from Mexico
Chamaedorea radicalis Martius from Mexico
Jubaea chilensis (Molina) Baillon from Chile
Nannorrhops ritchiana (Griffith) Aitchison from Afghanistan
Nannorrhops naudiniana Beccari, of unknown origin, and
Washingtonia filifera (Linden ex André) H.A. Wendland, from southern California and western Arizona although widely cultivated around the Mediterranean basin.

Amongst a comprehensive collection of species of the genus *Cupressus* found at the Villa Thuret are several notable rare species:

Cupressus chengiana Hu, from China
Cupressus dupreziana A. Camus, from the central Sahara
Cupressus gigantea Cheng & L.K. Fu, from China, and

Cupressus guadalupensis S. Watson, from the island of Guadeloupe. Amongst the *Eucalyptus* species mentioned as threatened in Australia, two species are cultivated at Villa Thuret:

Eucalyptus kartzoffiana Johnson & Blaxell
Eucalyptus risdonii Hooker f., which is found in a forest arboretum in the Massif de l'Estérel.

Other notable species are:

Docynia delavayi (Franchet) Schneider, (Rosaceae), from China
Notelaea excelsa Webber & Berthelot, (Oleaceae), from the Canary Islands
Olmediella betscheleriana (Goeppert) Loesener, (syn. *Licopolia syncephala* Rippa), (Flacourtiaceae), from Brazil.
Pittosporum coriaceum Aiton, (Pittosporaceae), from the Canary Islands and Madeira.

The majority of these species are represented at the Villa Thuret by just one or a very small number of specimens. This is insufficient to ensure successful genetic conservation of the species or sexual reproduction without the risk of hybridization. The garden's role in regard to these rare plants is therefore as a "shop window" where specimens are conserved and exhibited to the public. The vegetative propagation allows only for the distribution of plants to collectors or to horticulturists, without really participating in the conservation of the species.

However, two activities which are more markedly orientated towards conservation, are carried out by INRA in Antibes.

Jubaea chilensis (Molina) Baillon (Syn. *Jubaea spectabilis* H. B.K.)

Identified in 1815, this palm is mentioned as been given the IUCN Red Data Book category of vulnerable (V) on IUCN lists. In its natural environment in Chile, it was decimated by Man. It was collected for its very sweet sap, for the consumption of its terminal meristem (palm cabbage) and for the manufacture of ropes. Called "Glilla" in Chile or

Figure 1. *Jubaea chilensis* in the Jardin Botanique de la Villa Thuret

"Coquito" in Peru, the Chilean coconut palm grows spontaneously amongst dry high altitude bush (as mentioned by Charles Darwin in his account of the voyage of the "Beagle"), but in our gardens it has a more majestic look.

This very beautiful palm was introduced to France for the first time around 1850 by Charles Naudin to a place called Collioure in the eastern Pyrenees. Sadly, there are not many of them today, probably because they grow so slowly.

The specimens which were grown from the original seeds which Naudin gave to Villa Thuret between 1856 and 1860, are the parents of 9 specimens currently maintained in the park. Every year they produce between 4,000 and 6,000 seeds which are distributed to other botanic gardens via the international seed exchange scheme, or to nursery and seed merchants who create the demand. Germination is not easy. Once a seed-bed is mature, a few little plants appear after two or three months, and then a greater number after eight to ten months, and then others later still.

Of all the palms introduced to the Côte d'Azur the *Jubaea* is without any doubt the most majestic and imposing one. The trunk with a circumference that can grow to 3.5 metres, has a fairly smooth bark of pale grey colour. The leaves fall, leaving wide oval scars. The crown has 50 to 60 leaves about 4 to 5 metres long, green to blue-green in colour, forming a dense spherical shaped mass. Each leaf has numerous leaflets and regularly spaced along the entire length of the rachis. Their silvery underside gives a very characteristic appearance to this magnificent palm.

After flowering in June, bunches of fruits appear, and these resemble little coconuts, about the size of an apricot. This species resists cold weather very well. There was no damage noticed during the exceptional winters of 1956 and 1985, when the temperature fell to -10°C. However, the death of a few specimens of around 5 years of age was notified in 1870 in Montpellier, when a temperature of -18°C was experienced. Nevertheless, even then certain individuals survived.

Although conditions at the Villa Thuret Garden are sufficiently favourable for the cultivation of numerous palms,-which is sufficient to ensure that sexual reproduction can take place, their genetic variability is probably very reduced. All these individual plants are the progeny derived from the same genes. To be able to evaluate the production of seeds it would be necessary to introduce new gene lines, but until now it has proved impossible to obtain new seeds of Chilean origin. Similarly, information on the state of wild populations remains fairly difficult to obtain.

Cupressus dupreziana A. Camus

The family Cupressaceae has been known in Europe and California since early times and was pushed south during the Quaternary era. Of the three Mediterranean species, *Cupressus sempervirens* L. is the most widespread having been distributed in part by human use. On the other hand *Cupressus atlantica* Gaussen is strictly confined to the Moroccan Atlas mountains and *Cupressus dupreziana* A. Camus is confined to the central Sahara.

The story of *Cupressus dupreziana* is both dramatic and thrilling; it constitutes one of the most remarkable relicts of the plant world. For more than 100 years, scientists have been interested by the presence in the mountains of Tassili and Hoggar of an endemic cypress. Since 1860 the British ornithologist Tristan and the geographer Duveyrier had talked of the existence of a resinous plant in the centre of the Sahara. The local people of that region, the Touaregs, used its wood, calling it "Tarout", for the manufacture of domestic tools and instruments.

The camel captain Duprez had the privilege of being the first European to discover this tree in 1924. The species was described 1926.

Botanical characteristics

Although similar in some ways to *Cupressus sempervirens*, there are distinct differences which allow the two species to be distinguished from each other. In cultivation the young tree (of 10 to 15 years old) presents a neat and dense shape. The branches grow horizontally at first for about 15 cm. and then change their direction and grow vertically, thus making an elbow of a 90° right angle.

The end branchlets bend in the same direction and are disposed in a distal fashion. This specific characteristic allows for the identification of Duprez's cyprus in comparison with the *C. sempervirens*, on which the branches grow in all directions. The foliage is more or less green with a

Figure 2. *Cupressus dupreziana* in the Arboretum forestier, Massif de l'Esterel

tinge of blue-green, opposite and decussate. The scales contain resinous glands.

The solitary cone is ovoid and ellipsoid, and from 12 to 24 mm. long, of a pale matt brown colour, with 10 to 12 scales. The scales are indistinctly mucronate. Consequently, it is smaller than that of the *Cupressus sempervirens* and nor is it spherical, which distinguishes it from the *Cupressus atlantica*. Besides this it is characterized by a great deal of pollen.

In its natural environment, in old age it can reach great dimensions. The largest individuals can grow up to 18 m. with a trunk circumference at chest height of 7 m.

The shape of older trees is probably conical to spreading, but those from Tassili are very mutilated by man and the extreme climatic conditions.

Natural environment

The species is in fact distributed these days over a relictual area of 200 to 250 sq.km., situated in the central Sahara at Tassili n'Ajjer. Various missions have identified 153 individuals on the plateau of Edehi in l'Oued Tamerit, at an altitude of between 1,600 and 1,800 metres. These trees are all very old, and are not reproducing themselves. The species has disappeared from Hoggar where there are now only a few dead trunks to be found.

Protection and collection

In 1959, Professor Laferrere went across the Massif du Tassili and identified the living examples. He brought back a few seeds for French botanic gardens: the garden of La Tête d'Or in Lyon, the garden of the Faculté of Toulouse, the garden Les Cêdres at St. Jean Cap Ferrat, and the Villa Thuret at Antibes. The very poor germination of these only produced a few little plants, of which three are at the Villa Thuret. The three individuals were the start for local cultivation tests and vegetative multiplication (by cuttings and grafting).

Subsequently, other seeds were collected most notably during Professor Barry's mission. Mixed results were obtained from these cultivation tests. Nevertheless the species was distributed around the Mediterranean basin and in different countries with subtropical climates. In France, its commercialization for ornamental and forest use, had already started, and its propagation by vegetative means assured.

A first collection was gathered by F.A.O. at the Institut de Recherches Forestières in Tunisia. This collection was reproduced and saved to form the basis with which to constitute a small forest arboretum in France in the massif de l'Estérel. The 35 clones thus gathered together were complimented by 100 individuals which were the product of seeds from Tassili and which were gathered from identified trees (for example, No. 39 of Professor Barry's collection). In total, the cloned sample represented about 9% of the wild population and was complemented by more than a hundred seed progeny individuals with the aim of gathering together a collection which would be as representative as possible of current genetic variability.

This arboretum now has 285 samples which are beginning to fruit. The arboretum presents isolated forest conditions, and is hopefully sheltered from the contamination of external pollination because *Cupressus dupreziana* will hybridize with *Cupressus sempervirens*. The arboretum was conceived for the study of this species, and to act as a seed orchard as well as a means of propagating the species sexually.

Future plans for botanical surveys at the Villa Thuret include the following:

- the protection and conservation of the arboretum as a natural estate,
- the morphological and phenological study of the species *ex situ*,
- the study of vegetative propagation,
- verification of the viability of seeds obtained in the arboretum (germination tests and observation of morphological characteristics of young plants).

In this way the Jardin Villa Thuret hopes eventually to find the means of propagating the species and producing good quality plants, either by vegetative propagation of well documented clones, or by obtain seeds with wider genetic variability. The last of these remains less likely, given the relictual state of the wild populations. At the same time a team of pathologists at the Villa Thuret is planning to study the behaviour of the species in the presence of *Seiridium cardinale*, in order to closely examine clones which are tolerant to the fungus (J. Ponchet and C. Andreoli, pers. comm.).

Conclusion

The examples of *Jubaea chilensis* and of *Cupressus dupreziana* illustrate the research which has been carried out in the Jardin Botanique de la Villa Thuret for the conservation of plants. This research reflects well the new direction of botanic gardens for conservation, while at the same time respecting the specific interests of the garden, that is to say the introduction of exotic woody species as ornamentals or for forestry and perfecting techniques for their propagation.

The research carried out with *Cupressus dupreziana* serves to illustrate the criteria necessary for the successful rescue of an endangered species *ex situ*. They are:

- and exhaustive inventory of wild populations;
- correct sampling of the existing gene pool; and
- the use of appropriate propagation techniques.

However, although these activities are essential for the *ex situ* rescue of endangered species they should not regarded as a replacement for the protection of wild environments.

References

Andre, E. (1902). Les *Jubaea* (cocotiers du Chili). *Revue Horticole*, 8-11.

Arbez, M. (1987). *Les ressources génétiques forestières en France. Tome 1 : Les Conifères*. Paris INRA, B.R.G.

Barry, J.P., Belin, B., Celles, J.C., Dubost, D., Faurel, L. & Hethener, P. (1973). Essai de monographie du *Cupressus dupreziana* A. Camus; cyprès endemique du Tassili des Ajjer (Sahara Central). *Bull. Soc. Hist. Nat. Afrique N.*, 61(1-2), 95-178.

Bois, D. (1937). *Les plantes alimentaires chez tous les peuples et à travers les âges. Histoire, utilisation, culture*. Paris, Paul Lechevalier.

Camus, A. (1926). Un cyprès nouveau du Tassili. *Bull. Mus. Hist. Nat. Paris*, 32(1), 101.

Camus, A. (1926). Le *Cupressus dupreziana* A. Camus, cyprès nouveau du Tassili. *Bull. Soc. Dendrol. France*, 58, 39-44.

Duprez (Capitaine) (1926). Note du 27 février 1926. In Maire H.: Etudes sur la flore et la végétation du Sahara Central. *Mém. Soc. Hist. Afrique N.*, 3, *Mission du Hoggar*, 2, 49-50.

Franclet, A. (1967). Une méthode de greffage du *Cupressus dupreziana* sur *Cupressus sempervirens*. *Revue Forestière Franç.* 1, 5, 338-342.

Maire, R. (1944). Sur la fructification du *Cupressus dupreziana* A. Camus, du Tassil n'Ajjer. *Bull. Soc. Hist. Nat. Afrique N.*1, 35 (2), 12.

Naudin, C. (1894). Fructification du *Jubaea spectabilis* en France. *Revue des Sci. Nat. Appl. (Société nationale d'acclimatation)*, 22, 1-2.

Quezel, P. (1954). Contribution à la flore de l'Afrique du Nord.IV : Contribution à la flore du Hoggar. *Bull. Soc. Hist. Nat. Afrique N.*, 45 (1-2), 55-67.

Sahut, P. (1887). Lettre au président à propos de ses plantations de *Jubaea*. *Bulletin Soc. Nat. Acclim.*, 654-655.

Van Campo-Duplan, M. (1953). Recherches sur la phylogénie des Cupressacées d'après leurs grains de pollen. *Trav. Lab. Forest. Toulouse,* II sect. Tome II, 4(3), 20.

Résume

Le Jardin Botanique de la Villa Thuret à Antibes (France) a pour vocation l'introduction et l'acclimatation de plantes ligneuses d'origine subtropicale. Il joue en particulier un rôle de "vitrine" où sont présentées des espèces rares ou menacées dans leur aire naturelle. Parmi elles, *Jubaea chilensis* est représentée par une dizaine de sujets dont la fructification permet la récolte et l'expédition de plusieurs milliers de graines chaque année. Par ailleurs un long travail de collaboration a permis de rassembler 285 individus de *Cupressus dupreziana.* Ces arbres, issus de multiplication végétative ou sexuée, sont plantés en arboretum forestier pour y être étudiés *ex situ* et constituer un verger à graines.

Resumen

El Jardín Botánico de la "Villa Thuret" en Antibes (Francia) tiene como finalidad la introducción y la aclimatación de plantas leñosas de origen subtropical. En él se muestran especies raras o amenazadas en su entorno natural. Entre ellas, *Jubaea chilensis* está representada por una decena de individuos fértiles, permitiendo de este modo la recogida y distribución de varios miles de semillas annualmente. Por otra parte, gracias a un largo trabajo en colaboración se ha logrado reagrupar 285 individuos de *Cupressus dupreziana.* Estos árboles producidos bien asexualmente o sexualmente constituyen un "arboretum" forestal con dos objectivos: su estudio *ex situ* y ser un reservorio de semillas.

The potential rôle of the Rio de Janeiro Botanical Garden in conserving the Atlantic Forest of Brazil

Gustavo Martinelli

Jardim Botânico do Rio de Janeiro
Rio de Janeiro, Brazil

The Atlantic Forest is located in eastern Brazil and is amongst the most endangered forest in the world. In the past, it stretched over 1,000,000 sq. km., in a band 50 to 150 km. wide. At present, only 4% of the original forest remains, and in some regions there are only vestiges left.

Deforestation in Brazil has taken place over the past 400 years since early settlement by Europeans. The exploration system in the region was based in extraction, substituting the native forest with exotic cultures and the uncontrolled exploitation of timber, fibers and coal, and by the expansion of the cities themselves. The region is now the major population centre in Brazil and the highest industrial growth area.

The Atlantic Forest region presents a unique tropical and subtropical vegetation which includes:

- Coastal forest on the narrow low-lying coastal plain;
- Forest on the foothills of the mountain ranges that run parallel to the coast, including *Podocarpus* and *Araucaria* Forest;
- Forest on the slopes of the mountains themselves up to 1,800 m.;
- Montane grassland on upper slopes.

The most striking feature of the Atlantic Forest is its great floristic diversity and high rate of endemism.

The Rio de Janeiro Botanic Garden, created in 1808, is located in the State of Rio de Janeiro, one of the main centres of endemism of the

TROPICAL BOTANIC GARDENS: THEIR ROLE
IN CONVERVATION AND DEVELOPMENT
ISBN 0—12—346850—7

Atlantic Forest. It has a total area of 141 ha., of which 71 ha. are of native or regenerated forest.

The Botanical Garden has one of the largest herbaria in Latin America with almost 650,000 species, a fruit collection with 6,200 samples, a wood samples collection with 7,239 specimens and an arboretum with more than 7,000 live specimens in its collections. Of non-woody plant groups, notable collections of orchids and bromeliads are also maintained. Its library is one of the most complete in Latin America, with 34,000 books on botany. Its staff is composed of 54 researchers, 18 technicians, 74 gardeners and several students integrated with the scientific body. In 1988 we had 1 million visitors.

For many decades we have been focussing on preserving the Atlantic Forest, collaborating in the creation of conservation units such as national parks, biological reserves and environmental protection areas. We have also been developing research in these areas, collecting and organizing data, to improve the existing management plans.

We are promoting intensive fieldwork to identify new areas to be included in the conservation units system, since we are convinced that the existing protected areas are not sufficient to conserve the genetic resources of this great biological diversity.

I believe that one of the great difficulties for the effective conservation of the Atlantic Forest is the very scarce availability of information about the flora, preventing us from creating adequate management plans and effective control of these areas and also making environmental education of the population very difficult. Among the several activities being developed at the Botanic Garden, the Atlantic Forest Programme has been given priority and it is integrated with the work of the National Research Council of Brazil. The project has the following objectives:

1. Selection and demarcation of the remaining and most vital areas of Atlantic Forest in the State of Rio de Janeiro.

2. Structural characterization of the vegetation, in quality and quantity, of the selected areas, through a floristic and phytosociological methods.

3. Identification and characterization of endemic and endangered species.

4. Creation of a germplasm bank *ex situ* and *in situ*.

5. Characterization of the abiotic environment.
6. Creation of Atlantic Forest data bank.
7. Formation and training of human resources.
8. Planning and promotion of conservation units with government and non-government organizations.

This project, initiated in 1987, involved last year 9,000 man-hours in the field. At the present we are tabulating and organizing the data obtained in the first area under study. This data will justify the creation of a research station there which will be brought in to being with the participation of the local community.

The area of action of the Atlantic Forest Programme is located in the headwaters of the Flores and Macae rivers, in the Country of Nova Friburgo and adjacent areas in the State of Rio de Janeiro, an area covering 14,000 ha, between 700 and 1,500 m. altitude.

Our methodology uses as parametre a minimal area of 3 ha., noting individuals with a 2.5 cm. minimal diameter, lianas and vascular epiphytes. In 1 ha. surveyed, 3,820 individuals of 313 species were measured and collected, including 4 new species.

Another project related to the Atlantic Forest is being developed by the Botanic Garden of Rio de Janeiro, with the support of the World Wide Fund for Nature (WWF). This project surveys the floristic resources of the upper grassland which occur on the top of high mountains in the Rio de Janeiro State, above the altitudinal limits of the forest. The preliminary results show 84 endemic species, 30% of which are endemics of restricted distribution. Most of them are endangered, due to their small population size and to human interference with their habitats. In this area 4 new species were discovered. Based on the preliminary results obtained, an Environmental Protection Area was created and another one is being proposed.

I need to admit our lack of experience in developing new conservation strategies. We feel that it is very important to exchange ideas and experiences with botanic gardens all over the world. We are interested in obtaining effective help for the execution of the Atlantic Forest Programme, particularly in funding, strategies and training of human resources.

In a country of continental dimensions such as Brazil, we have a huge task at our hands. We need to integrate ourselves with the international organizations take into account world-wide conservation strategies, in which botanic gardens play a fundamental rôle.

Part 3

Tropical Botanic Gardens: Their Rôle in *in situ* Conservation

Indigenous and alien plants in the Galápagos Islands: problems of conservation and development

Ole Hamann

Botanic Garden, University of Copenhagen, Copenhagen, Denmark

Summary

The unique flora and vegetation of the Galápagos Islands, Ecuador, are threatened by land use practices and by invasive alien animals and plants. The current conservation measures to meet these threats, the future prospects for conserving indigenous plants, and for ensuring their rational utilization are discussed.

Introduction

The Galápagos Islands are located about 1,000 km. west of continental Ecuador. The archipelago spreads over 45,000 sq. km. of ocean and consists of 13 major islands and numerous smaller islets and rocks, in total about 8,000 sq. km. There is a characteristic variation in habitats and local climatic and ecological conditions, largely shaped and determined by the ocean currents, the prevailing winds and the active, volcanic nature of the islands.

The Galápagos Islands are true oceanic islands of volcanic origin and are between 0.7 to perhaps 4.5 million years old (Simkin, 1984; Geist et al., 1985). They were first discovered in 1535, but were not colonized until 1832, when the archipelago formally was annexed by Ecuador and a small, permanent colony established (Perry, 1984).

TROPICAL BOTANIC GARDENS: THEIR ROLE
IN CONVERVATION AND DEVELOPMENT
ISBN 0–12–346850–7

The flora and fauna are characterized by a high degree of endemism, having evolved in isolation and free from human impact until a few centuries ago. The biological isolation of the islands and their ecosystems ended with the arrival of buccaneers in the 17th century; their impact was presumably not great, but later on, in the 18th and 19th century, whalers and sealers took a heavy toll on the islands' populations of tortoises and fur seals. When permanent settlements were established from 1832, the impact of man on the islands increased, but, until 1950, the resident population remained below one thousand. However, since then the number of resident inhabitants has grown, especially during the last decades: from 1974 to 1988 the number rose from 4,000 to over 10,000. This rise is mainly a result of the development of tourism in the islands, which has been accompanied by an influx of immigrants from mainland Ecuador.

Already in 1934 the Ecuadorian Government realized that many unique animals and plants were threatened with extinction and the first laws to protect the indigenous wildlife were passed; but it was first in 1959 that practical conservation measures were initiated when about 97% of the land surface of the archipelago was declared a National Park. In 1986 the surrounding waters, some 80,000 sq. km., were declared a Marine Reserve, and in 1990 all Ecuadorian national waters, including the Galápagos Archipelago, were declared a whale sanctuary.

The conservation problems related to plants and vegetation have for long been recognized by the Charles Darwin Foundation for the Galápagos Isles (CDF) and the Galápagos National Park Service (GNPS). The CDF is an international, non-profit organization established to help promoting science and conservation in the Galápagos Islands. The CDF has an agreement with the Ecuadorian Government to operate the Charles Darwin Research Station (CDRS) on Santa Cruz Island, and to advise the government on all matters related to science and conservation in the islands. The principal counterpart of the research station is the GNPS, which is the national, Ecuadorian institution that manages the Galápagos National Park.

Although conservation legislation has been enforced and practical conservation measures implemented since the creation of the CDF and the NPGS in 1959 (Corley Smith, 1990), the Galápagos Islands are faced with grave conservation problems. Since the time of the buccaneers and whalers humans have exploited the natural resources of Galápagos, and indigenous ecosystems have been altered and modified. A large number of alien organisms, plants and animals, has been introduced, and many have turned out to be aggressive invaders. Furthermore, the present boom in tourism puts increasing pressures on the islands with respect to both to the carrying capacity of the natural resources, and the infrastructural and institutional management capacity. There is now strong indication that the limits for rational utilization of the natural resources, set by the very nature of the islands, are exceeded, and the consequence may be rapid deterioration of ecosystems. In particular, the plant diversity of the islands is today threatened by habitat fragmentation, invasive alien species (e.g. feral mammals and introduced aggressive plants), and over-exploitation of the limited plant resources by a growing human population.

The indigenous flora and vegetation

a) The flora

Since the famous visit of Charles Darwin to the Galápagos in 1835 the islands have been known to have a very unique environment, caused by the conjunction of a number of peculiar factors. Hardly any other archipelago in the world is at once so extensive, so isolated, so simple in its ecosystems and so little altered by man. In correspondance with the isolation of the archipelago, the volcanic origin and the constant presence of open pioneer habitats, the Galápagos flora contains a high number of endemic taxa.

According to Lawesson (1990a) the native flora of Galápagos comprises about 596 taxa, of which 224 are endemic and 372 are indigenous, non-endemic ("natives"). The endemism is thus about 38%, including subspecific taxa. Adsersen (1989) lists a total of 604 taxa as indigenous, also including subspecific taxa. On the species level, there are about 229 endemics among a total of 541 species, i.e. about 42% endemism (Loope et al., 1988). It is noteworthy that there are seven endemic genera among the flowering plants and that the endemism in this group of plants is as high as 51%, while among ferns and allies it is not more than 7% (Loope et al., 1988). Considering the age of the archipelago (0.7-4.5 million years) the percentage endemism is very high.

Plant names in this paper are in accordance with Lawesson et al., 1987).

About 260 introduced, non-native plants are now recorded from the archipelago and the number is steadily increasing (Lawesson, 1990b). Many introduced plants have become naturalized, such as pantropical weeds, but about 10 to 17 of the introduced plant species are considered to be aggressive invaders (Hamann, 1984; Lawesson, 1990b).

Of the 604 indigenous plant taxa listed by Adsersen (1989) about 24% were termed "rare", a term employed to correspond loosely, but not strictly, to the rare or threatened categories used by IUCN (Anon., 1980). Lawesson (1990a) evaluated the Galápagos flora in strict accordance with the IUCN Red Data Book categories and came to the result that 2 endemic species are Extinct, 20 endemic or indigenous taxa are Endangered, while 16 endemic or indigenous taxa seem to be Vulnerable. Among these threatened taxa are representatives of the endemic genera *Scalesia* (Fig. 1), *Lecocarpus* and *Darwiniothamnus*, all members of the Asteraceae.

Figure 1. *Scalesia incisa*, grown in the Botanic Garden, University of Copenhagen. The genus *Scalesia* is endemic to Galápagos and comprises 21 taxa. *S. incisa* is restricted to the northern part of San Cristóbal, where in some parts of its distribution range it is threatened by feral goats.

In contrast with Adsersen (1989), Lawesson listed only 33 native taxa as Rare, while 370 taxa were placed in the non threatened category (nt). As Lawesson pointed out, many endemic taxa in Galápagos are rare simply because by definition they are restricted geographically and because many endemics tend to have very small populations. Even so, many Galápagos endemics occur in such numbers and over such areas that they probably never will become threatened, e.g. *Pisonia floribunda*, *Psidium galapageium*, *Scalesia affinis* and *Castela galapageia*.

b) The vegetation

The Galápagos vegetation has been dealt with in many studies and a number of different vegetation types and plant communities has been recognized and described (e.g. van der Werff, 1978, 1990; Hamann, 1981; Eliasson, 1984; Atkinson, 1990; Itow, 1990). However, for convenience, the broad concept of vegetation zones as described from Santa Cruz island by Wiggins & Porter (1971) is often referred to, viz. littoral, dry, transition, *Scalesia*, *Miconia* (Fig. 2), and fern-sedge zones. Actually, a Brown zone should be included, since this particular type of vegetation in former times occurred between the *Scalesia* and the *Miconia* zone, at least on Santa Cruz island (Bowman, 1961).

These vegetation zones relate in particular to altitude and rainfall. With respect to threatened vegetation types and plant communities, Hamann (1984) made an assessment of the endangered vegetation types of the archipelago. He concluded that first of all it was the more mesic vegetation types that were at risk, such as the fern-sedge zone vegetation and the *Miconia* zone on Santa Cruz and San Cristóbal, the Brown zone on Santa Cruz, and much of the *Scalesia* zone on San Salvador, San Cristóbal, Santa Maria and Santa Cruz. On the other hand, the arid vegetation types, in the widest sense, were for many reasons considered to be the most robust of Galápagos vegetation types, in addition to being those covering the largest areas in the archipelago. However, it is noteworthy that the analysis also showed that the majority of Galápagos ecosystems simply is characterized by a very high proportion of endemic species (which often dominate the plant communities, for example), so the plant communities and the vegetation types themselves can be considered endemic and therefore literally unique (Hamann, 1984, 1981). Thus when Lawesson (1990a) discussed the vegetation of the 21

Figure 2. *Miconia robinsoniana*, the endemic Melastomataceae. The *Miconia* shrub type of vegetation has almost completely been destroyed by clearing and cattle grazing on San Cristóbal and is threatened by invasive plants on Santa Cruz.

key sites where the Endangered and Vulnerable taxa occur, he in fact also indicated which parts of endemic, unique vegetation types should be given conservation priority. Most of the key sites listed by Lawesson are located on the four inhabited islands, in correspondance with the many adverse factors related to the presence of human settlements. Only 4 sites on 4 uninhabited islands were considered of priority for plant conservation reasons by Lawesson (1990a). In addition to key sites, Lawesson also identified some vegetation types and plant communities as being threatened, as for example the Brown zone mentioned above, the *Cyathea* thickets on Santa Cruz and the *Sphagnum* bogs and fens found in the highlands of Santa Cruz and San Cristóbal islands.

The threats

a) Land use practices

Adsersen (1989) estimated that 54% of the 144 taxa, which he classified as "rare", are threatened by land use practices, 40% by introduced animals, and 19% by introduced, invasive plants. Adsersen listed and discussed both the natural and the man-induced threats to the "rare" plants, but in general landuse practices and introduced plants were identified as causing the worst problems on the inhabited islands, while feral mammals were considered the worst on some of the uninhabited islands.

Lawesson (1990a) considered in detail the 36 taxa, which he classified as belonging to the categories Endangered and Vulnerable according to the IUCN classification, and came to the same general result as Adsersen (1989) and Hamann (1984), viz. that the main threats to the Galápagos plants are introduced animals and plants, and human impact. Correspondingly, almost all Endangered species seem to be concentrated on the inhabited islands of Floreana, San Cristóbal, Santa Cruz and Southern Isabela, and on the uninhabited island of Santiago, which has large populations of feral goats and pigs.

The landuse practices particularly present a threat to the plants and vegetation types of the humid highlands on the southern slopes of the larger islands (Hamann, 1984; Adsersen, 1989). These humid areas are relatively small and are the only land appropriate for agricultural use in the islands; consequently they were already early in colonization taken over for agricultural purposes and were therefore not included in the National Park at it's creation in 1959. Although the present agricultural areas only comprise about 6% of the total area of the Galápagos, they comprise roughly half the area originally covered by humid vegetation types (Adsersen, 1989). The remaining half of humid vegetation types and habitats, which thus is part of the National Park and therefore under protection, is nevertheless under growing threat. Firstly because introduced animals and plants are able to spread into the National Park from the colonized zones, and secondly because of the growing human population, which exerts an increasing pressure on the resources of the Park with respect to an actual, but illegal, or intended exploitation of, for example, timber and grazing lands for cattle.

b) Invasive animals

The main introduced and invasive organisms in the Galápagos have recently been discussed by a number of researchers (e.g. Hamann, 1984; Hoeck, 1984; Brockie et al., 1988; Loope et al., 1988; Schofield, 1989; Lawesson & Ortiz, 1990). A global perspective on wildlife conservation and the invasion of nature reserves by introduced species was given by Macdonald et al. (1989). In the Galápagos, the flora and vegetation are evidently threatened by feral mammals and by invasive, exotic plants (Hamann, 1984). The distribution of introduced mammals in the Galápagos (Table 1) shows why Endangered taxa and unique vegetation types are most at risk on the inhabited islands: all these islands have a large contingent of introduced mammals, which have spread outside the colonized zones into the National Park areas. Even on uninhabited islands like Santiago, Pinta and Pinzon, introduced mammals are, or have been, causing grave problems. The general effect of particularly feral mammals such as goats and pigs is an opening up of closed native vegetation, so the composition and structure of the vegetation is changed. This has been amply documented on e.g. Santiago (Calvopina & de Vries, 1975; de Vries and Calvopina, 1977) and on Pinta (Hamann, 1979; 1984). The alterations on Santiago, where perhaps as many as 100,000 feral goats and several thousand feral pigs are found, are so profound that erosion has started in places, and the biochemical cycling of the ecosystems may have been altered. The *Scalesia* zone vegetation, originally dominated by *Scalesia pedunculata*, is almost completely gone from Santiago, and the local endemics *Scalesia atractyloides* (both varieties) and *Scalesia stewartii* are now listed as Endangered and Vulnerable, respectively (Lawesson, 1990a).

c) Invasive plants

Of the about 260 introduced plants now occurring in the Galápagos, those that turn out to be aggressive invaders in the long term may present an even more serious conservation problem than the introduced mammals. Introduced plants are recorded from most islands (Lawesson, 1990b), but so far the invasive plants are mainly restricted to inhabited islands. However, many introduced species are spreading out into uninhabitated areas of National Park on these islands, and one or two invasive taxa have also become established on uninhabited islands. Table 2 lists

Table 1. Distribution of introduced mammals in the Galápagos.

Islands and areas	Goats	Pigs	Donkeys	Cattle	Horses	Sheep	Cats	Dogs	Black rats	Brown rats	House mice
Santa Cruz	*	*	*	x			*	*	*	*	*
San Cristóbal	*	*	*	x			*	*	*	*	*
Floreana	*	*	*	x			*	*	*		*
Isabela, Volcan Sierra Negra	*	*	*	*	*	*	*	*	*		*
Isabela, Volcan Cerro Azul	*	*	*	*			*	*	*		*
Isabela, Volcan Alcedo	*		*				*		*		?
Isabela, Volcan Darwin							*		*		?
Isabela, Volcan Wolf							*		*		?
San Salvador	*	*	*						*		
Pinzon									*		
Pinta	x										
Marchena	x										
Espanola	x										
Santa Fe	x										
Rabida	x										
Genovesa											
Fernandina											

* = present; ? = maybe present; X = exterminated by the GNPS.

the main invasive plants in the Galápagos; of these, the most dangerous threats to the native vegetation are *Psidium guajava*, *Cinchona succirubra*, *Lantana camara*, *Rubus* sp., *Pennisetum purpureum*, and perhaps *Passiflora edulis* and *P. ligularis*.

Psidium guajava was probably introduced to the islands around 1858 and is today the alien species with the largest extension in the archipelago; during the last 40 years or so, *Psidium guajava* has spread on the four inhabited islands both in the colonized zones and outside. On South Isabela alone it is estimated to cover some 20,000 hectares and on San

Table 2. Main invasive alien plants in the Galápagos.

Agavaceae:	
Furcraea cubensis	S. Cruz; S. Cris.; Flor.; S. Negra
Boraginaceae:	
Cordia alliodora	S. Cruz
Crassulaceae:	
Kalanchoe pinnata	S. Cruz; S. Cris.; Flor.; S. Negra
Lauraceae:	
Persea americana	S. Cruz; S. Cris.; Flor.; S. Negra; S. Sal.
Meliaceae:	
Cedrela odorata	S. Cruz; S. Cris.; Flor.
Myrtaceae:	
Eugenia jambos	S. Cruz; S. Cris.; Flor.; S. Negra
Psidium guajava	S. Cruz; S. Cris.; Flor.; S. Negra; C. Azul
Passifloraceae:	
Passiflora edulis	S. Cruz; Flor.
P. ligularis	S. Cruz
P. quadrangularis	S. Cruz; S. Cris.; Flor.; S. Negra
Poaceae:	
Axonopus compressus	S. Cris.; Flor.; S. Negra; C. Azul
Digitaria decumbens	S. Cruz; S. Negra
Pennisetum purpureum	S. Cruz; S. Cris.; Flor.; S. Negra
Rosaceae:	
Rubus sp.	S. Cruz; S. Cris.
Rubiaceae:	
Cinchona succirubra	S. Cruz
Rutaceae:	
Citrus limetta	S. Cruz; S. Cris.; Flor.
Solanaceae:	
Solanum sp.	S. Cruz
Verbenaceae:	
Lantana camara	S. Cruz; S. Cris.; Flor.

S. Cruz = Santa Cruz; S. Cris. = San Cristóbal; Flor. = Floreana; S. Negra = Volcan Sierra Negra, Isabela; C. Azul = Volcan Cerro Azul, Isabela; S. Sal. = San Salvador.
Data from: Hamann, 1984; Lawesson et al., 1987; Macdonald et al., 1988; Schofield, 1989; Lawesson & Ortiz, 1990; and personal observation.

Cristóbal some 12,000 hectares (Lawesson, 1990b). *Psidium guajava* has been able to invade native vegetation, often helped by the presence and activities of domestic or feral mammals, who open the native vegetation and spread the seeds. The species thus present a number of conservation problems: it alters the composition and structure of native vegetation types, prevents recruitment of native plant species, interacts with native and introduced animal species in a way that may favour further establishment and spread of introduced species (e.g. rats and pigs), and it probably changes the fire regime (as witnessed during the great fire in a 200 sq. km. area on South Isabela in 1985, cf. Nowak et. al., 1990).

Another major introduced plant problem is presented by *Cinchona succirubra* on Santa Cruz island (Hamann, 1974; Macdonald et al., 1988). This species was first introduced to Santa Cruz in 1946. In 1972 it occurred scattered in the highlands (National Park area), but by 1986 it was estimated to cover about 4,300 hectares in the highlands (Ortiz & Lawesson, 1987). *Cinchona* occurs throughout the colonized zone at lower altitude, although not in large numbers; from there the light, wind-borne seeds are easily dispersed upwards to the highlands, where the *Cinchona* trees have been able to establish themselves extensively in the *Miconia* and fern-sedge vegetation types, and where to a large degree they are outcompeting the native species. Considering that the highlands of Santa Cruz probably originally were treeless (Howell, 1942; Hamann 1984), the whole aspect of this part of the island is being altered by the invasion of *Cinchona*.

The recent spread of *Lantana camara* and *Rubus* sp. have shown that certain species are particularly aggressive and may take over large areas within a short span of time. *Lantana* was introduced as an ornamental to the island of Floreana in 1938 (Lawesson & Ortiz, 1990), but during the last decade it has spread explosively in the remnants of the *Scalesia* zone vegetation on this island. Today it covers some 2,000 hectares, and in addition to displacing native plants and vegetation, its vigorous growth prevents the endangered Hawaiian Petrel from easy access to the nesting area. Thus, *Lantana camara* may contribute to the global extinction of an Endangered bird. The recent spread of *Rubus* sp. in the Galápagos has been even more rapid than that of *Lantana* and *Cinchona*. The *Rubus* sp. was introduced to San Cristóbal in 1983 for its fruits, but already by 1986 it had taken over more than 100 hectares of the highlands and in 1989 it covered probably more than 200 hectares (pers. obs.). It is also now present on Santa Cruz island. The most obvious effect of *Rubus* is simply that it smothers native plants, such as the remaining shrubs of *Miconia robinsoniana* still occurring in the highlands of San Cristóbal. The recent spread of such species as *Lantana* and *Rubus* were facilitated by the previous opening up or destruction of the native vegetation by domestic or feral mammals, and then assisted by birds (particularly Darwin's Finches) dispersing the seeds.

Two more introduced species, *Passiflora edulis* and *P. ligularis* may turn out to be very troublesome in a few years, according to Lawesson & Ortiz (1990). Species of *Passiflora* are recorded as aggressive invaders in other parts of the world, e.g Hawaii, and the recent spread of the two species on Santa Cruz island in the Galápagos is alarming.

Combined effects of land use practices and invasive organisms

Oceanic island ecosystems are vulnerable to biological invasions, and particularly to certain types of invasions related to and caused by man. For example, in the Galápagos, giant tortoises and land iguanas are the largest naturally occurring herbivores, so the introduction and spread of herbivorous mammals completely altered the ecosystems and placed tremendous pressure on the native plant species. At the same time, many introduced plant species evolved under high grazing and browsing pressure from herbivorous mammals and are therefore well adapted to such pressure. Consequently, the combined effect of introducing into the Galápagos both herbivorous mammals and introduced plants adapted to grazing can be disastrous for the native plants. Furthermore, not only are endemic and native species and plant communities threatened with extinction, but the combined effects of land use practices and invasive species may alter the biochemical cycling in ecosystems, change the fire regimen, and prevent recruitment of native plants (but further the spread of other, mainly introduced species). The possibility for genetic effects of introduced species should also be considered, e.g. when *Lantana camara* moves into areas occupied by the endemic *Lantana peduncularis*. Another important point is that although it often is possible to predict which plants may become invasive, because of their evolutionary history, adaptations and attributes (e.g. *Rubus*), it is not always possible: the introduced *Cedrela odorata* in the Galápagos gave no cause for alarm during the first 25 to 30 years of its presence in the islands; but during the last 5 years an unexpected and sudden spread of the species into native vegetation has taken place (Lawesson & Ortiz, 1990).

Thus, the rather detailed knowledge now available on the Galápagos flora and vegetation and on the threats to the botanical diversity of the archipelago shows that even though the nature of the Galápagos Islands is in a relatively good state, compared to many other tropical areas and islands, there is an urgent need for conservation measures directed specifically towards plants.

The rôle of plants in local economy and development

a) The zones of the National Park.

For a major part of the Galápagos archipelago the plant "resources" are the native plants and vegetation types, which are only exploited indirectly, and which are given high priority with respect to management of the National Park. The Master Plan for the Galápagos National Park (Anon., 1974) divided the Galápagos archipelago into zones in order to provide a basic management tool for the GNPS. In correspondance with the intended use of the land, the following zones were recognized: Primitive-Scientific, Primitive, Extensive Use, Intensive Use, Special Use, and Colonized zone.

The Primitive-Scientific zone includes those areas that are essentially free of introduced species (e.g. Fernandina island and Volcan Darwin and Volcan Wolf on northern Isabela), and which require the highest order of protection to ensure their continued ecological integrity. This means that nothing in the way of human use is permitted if it intrudes or in any way damages or alters the natural scene. Access is restricted and is generally only given for scientific purposes. The Primitive, Extensive Use and Intensive Use zones are managed first of all according to their estimated carrying capacity with respect to visitors. The Primitive zone is by far the largest in the Park and includes extensive areas (e.g. Santiago island, Volcan Alcedo on Isabela etc.) which, although somewhat ecologically altered by introduced species, should retain their primitive character. Access is regulated so that the zone serves as a buffer between the Primitive-Scientific Zone and more accessible zones. The Extensive and Intensive Use zones contain areas of outstanding features of visitor interest (e.g. visitor trail to the east rim of Volcan Alcedo, Isabela, and

South Plaza and Bartolomé islands), which are estimated to be able to take light or heavy visitor loads, respectively.

However, the Special Use and Colonized zones are of particular interest with respect to the direct use of local plant resources. The Special Use zone includes substantially altered lands which are near the settlements (e.g. on the south slopes of Sierra Negra, Isabela and on Santa Cruz and San Cristóbal islands) and which, although in need of careful management, are not of the same quality as the rest of the Park. In this Special Use zone the residents may, with permits, collect firewood, log timber and extract wood on a limited basis, and collect introduced fruits.

Thus, the flora and vegetation serves many purposes of importance in the local economy and development, ranging from a basic resource for science and tourism to a direct supply of usable wood and timber.

b) The indigenous hardwood trees

There are only are few species of indigenous hardwood trees in the Galápagos. The main species are the native *Piscidia carthagenensis* and *Hippomane mancinella* and the endemic *Psidium galapageium*. It is in particular *Piscidia* that has been exploited for its excellent wood qualities, which makes it extremely useful for construction purposes (houses, boats, furniture etc.) and for durable fenceposts. Some 20 years ago, only *Piscidia* and *Hippomane* were used to any extent, and the extraction from the Special Use zones close to the settled areas was fairly limited, not the least because permits were only given for extraction of dead trees. Occasionally, some other woody species with much lighter wood, such as *Scalesia pedunculata*, were also used, e.g. for construction purposes. However, studies by the CDRS have clearly shown that the exploitation rate today is far greater than the natural regeneration rate in the National Park (Ramirez, 1988); the diminishing quantities of usable specimens of *Piscidia* and *Hippomane* has then recently lead to an increasing exploitation of *Psidium galapageium*.

The rapid growth of the human population in the Galápagos has accelerated the direct exploitation of hardwood trees because the much larger human population simply needs more wood resources for various purposes, but also because the general increase in prosperity of the resident population (which is a result of the increase in tourism) has changed the pattern of use of local tree resources. Many new houses have been constructed in cement or stone, many new boats have been built and many old fishing boats have been rebuilt to accommodate tourists, all of which has lead to a growing demand for timber of large size and good quality.

Plants in the colonized zones

The colonization of the Galápagos islands radically changed the humid, highland areas in particular. Land was cleared for pastures, crop fields and fruit plantations, and many animals and plants were introduced to the islands (tables 1 & 2). Today the four colonized islands, Floreana, Isabela, San Cristóbal and Santa Cruz, have somewhat different sets of problems related to plants and to management of plant resources, in agreement with their histroy of colonization (Perry, 1984; Hickman, 1985). For example, the occurrence and extent of distribution of introduced plant species are different on the four islands (cf. table 2). However, it is noteworthy that in general the introduced plants create as many problems in the settled, agricultural areas as they do in neighbouring National Park areas. For example, the spread of *Psidium guajava* in land set aside for pasture in the Colonized zone is a very real problem for many farmers, and such species as *Rubus* may well take over large parts of agricultural land, as well as native vegetation in the National Park. Furthermore, most problems related to introduced organisms have, naturally, their origin in the colonized zone, from where those species that are aggressive and invasive are able to spread out. Thus the GNPS has recognized the need for developing agroforestry programmes, which aim at integrating the conservation management of the National Park with the rational use of the land in the colonized zone.

Watersheds

In the Galápagos water is a very scarce resource; in the highlands of the larger islands, a major part of the precipitation comes in the form of fog condensation (Hamann, 1979b). The importance of condensation is illustrated both from uninhabited and inhabited islands. On the uninhabited island of Santiago, the highland vegetation has largely been destroyed by the thousands of feral goats living there. Without dense vegetation of trees and shrubs to condense moisture from the humid air, the highlands of Santiago are becoming increasingly arid (de Vries & Calvopina, 1977).

The inhabited San Cristóbal is the only island having permanent streams and a permanent fresh water lake. The highland region of San Cristóbal is cool and humid, and fog condensation is an extremely important feature. The natural vegetation comprises *Scalesia pedunculata* forest, *Miconia* shrub (Fig. 2), *Cyathea* tickets and *Sphagnum* bogs and fens. However, in large areas of the highlands, which are not included in the National Park, *Scalesia* forests have been cut down, and *Miconia* shrub and *Cyathea* thickets have been destroyed by extensive grazing of cattle and goats. Instead *Psidium guajava* and *Rubus* sp. have moved into the area. As a result of the removal of the original, dense vegetation around the many small creeks in the area a decrease in fog condensation and a reduction in the water flow has been recorded (Lawesson & Estupinan, 1987).

Current conservation measures of importance to plants

Conservation programmes undertaken by the CDRS and the GNPS during the last 30 years have successfully ensured the survival of many unique and endemic animal species, such as the famous giant tortoises and the land iguanas. Furthermore, the detailed zoning system, regulating the kind of use permitted in the different islands and areas, ensures that large parts of the archipelago are totally protected and remain relatively pristine.

Introduced mammals have long been recognized as the most immediate threat to the Galápagos ecosystems, and in particular to indigenous wildlife. Consequently, eradication or control of introduced mammals (cf. table 1) has for long been a major part of the programmes of the GNPS and the CDRS. The goat eradication programme in particular has been of great importance for the survival of plants and vegetation types in the archipelago. The complete elimination of feral goats has been achieved on a number of small and medium sized islands (cf. table 1), which has resulted in notable or even spectacular recovery of vegetation on such islands as Santa Fe and Pinta, respectively (Hamann 1979a,

1985; Whelan & Hamann, 1989). Ongoing campaigns against goats and pigs on Santiago and against goats and donkeys on Volcan Alcedo, Isabela island, are of vital importance for the survival of the plants there. On the large island of Santiago the destruction caused by goats and pigs is so profound that it is very doubtful whether the vegetation will be able to recover to its original state, even if the feral mammals can be completely eliminated. However, very small remnants of the various native vegetation types on Santiago have been fenced to keep out feral mammals, and such fenced plots are at present serving as small, living seed banks. At the same time, they clearly demonstrate the devastating effect that goats have had on the native vegetation (Calvopina & de Vries,1975; de Vries & Calvopina, 1977).

During recent years, the GNPS and the CDRS have started campaigns against some of the introduced and invasive plants, e.g. *Cinchona* and *Psidium guajava*. On Santa Cruz the wardens of the GNPS clean annually, by manual weeding and uprooting, some 1,000 hectares of *Cinchona*-infested highland in the National Park (Macdonald et al., 1988). Several trials have been conducted on eradication of *Psidium guajava* with the help of arboricides and other means, but so far no efficient method has been devised (Utreras 1987). In general, the problems caused by the invasive alien plant species have so far exceeded the capacity of the GNPS with respect to manpower and financial resources, but in the long term, the aim is to combine the elimination of invasive species in both the colonized zones and in the National Park with the cultivation of indigenous or introduced, non-invasive hardwood trees (Cifuentes, 1990).

In order to meet the demand for good quality wood and timber for construction purposes, poles and fenceposts etc. and to diminish the pressure on the native trees, the GNPS and the CDRS have for some years now carried out a pilot project on the cultivation of indigenous and introduced, non-invasive hardwood trees. A small nursery has been established on Santa Cruz, and local farmers have provided land for more extensive cultivation trials. A number of species are being investigated with respect to germination, growth and survival under Galápagos conditions, and so far it seems that the introduced *Tectona grandis* is the most promising. But even the indigenous *Hippomane mancinella* and *Piscidia carthagenensis* may be possible candidates for future plantations in the islands.

Related to these activities are those concerned with rehabilitation of the small watersheds on San Cristóbal island, where the local authorities and the CDRS have initiated a reforestation project aimed at (re)creating the ecological conditions around the creeks in order to guarantee an adequate supply of fresh water, even during droughts (Lawesson & Estupinan, 1987). The endemic species *Scalesia pedunculata* and *Miconia robinsoniana*, and the introduced (but presumably non-invasive) *Juglans neotropica* are being investigated with respect to seed germination and survival, but in spite of a promising start (cf. Lawesson & Estupinan, 1987), the project has met with several difficulties; for example, it turned out to be difficult to germinate *Miconia* seeds, and neither *Scalesia* nor *Juglans* did well at the experimental site, probably because the site was placed at an altitude which is above the optimum for the species (and therefore too cool). However, the natural regeneration of *Miconia* around the creeks seems to be good, and propagation of *Miconia* from cuttings has proved to be possible. So if the local authorities and land owners agree to fence off the creeks to keep cattle out, the prospects for being able to reforest the areas immediately around the fresh water streams with *Miconia* seem promising.

Future prospects

a) A comprehensive plant conservation programme

In spite of the many conservation achievements in the Galápagos, so far it has not been possible to mount efforts to ensure conservation of the plant diversity of the islands comparable to those implemented for conserving the animal diversity. A main reason for this is that the problems facing the GNPS and the CDRS in trying to conserve the plant diversity of the Galápagos Islands are very much interrelated. Solutions therefore have to comprise elements of strict protection, management of native plants and vegetation, and management of introduced and cultivated species, i.e. management of plants and vegetation outside the National Park areas.

In 1987, a major step forwards was taken when the CDF/CDRS together with the GNPS organized an international workshop in the Galápagos aimed at assessing how the disciplines of the botanical sciences and the knowledge gathered during many years of research in the archipelago could be translated into practical conservation management. Based on the recommendations of the workshop (Lawesson et al., 1990) the CDRS then formulated a comprehensive plant conservation programme for the Galápagos Islands which hopefully will be implemented during the coming years.

This comprehensive plant conservation programme consists of an initial 3-year phase considered the minimum necessary to halt the present negative trends, and a long term phase, which focuses on changing the current agricultural practices into more rational agroforestry schemes. The main short-term objectives and expected results are:

1) to protect threatened native vegetation types by fencing against feral mammals, and thereby preserving selected areas of native ecosystems as living seed banks;

2) to identify the most aggressive alien plants and their distribution in the National Park where they threaten native ecosystems, as well as in cultivated farmland where these species first became established, thereby getting an overview and assessment of the problems posed by the aggressive alien plants in the islands;

3) to develop safe, effective methods of control and eradication for introduced plants and to provide advice and help to farmers to control pest plants inside agricultural areas, thereby permitting more effecient land use; and

4) to assess the natural timber resources and develop and promote alternative native and non-native timber sources, thereby providing sustainable sources of timber and wood, resulting in reduced pressure on National Park resources.

The plant conservation programme will comprise training of GNPS personnel in plant conservation methods and techniques, implementation of quarantine measures to limit the inflow of introduced organisms from the mainland, and generation of increased awareness of the value of the natural plant resources and their management among the local population.

b) The *ex situ* conservation element

One important part in the long term plant conservation efforts in the Galápagos Islands is the development of an *ex situ* programme for threatened plants, to complement the ongoing and planned *in situ* activities in the archipelago. Following the recommendations made by the 1987 workshop (Lawesson et al., 1990), a programme for *ex situ* conservation has been proposed by the University of Copenhagen, Denmark. The aim is to carefully select, together with the GNPS and the CDRS, target species for *ex situ* conservation, and conserve them in the new seed and genebank at the Botanic Garden, University of Copenhagen. As mentioned by Lawesson (1990a) some of the Endangered Galápagos taxa are close to extinction and occur in very low numbers. For example, some of the *Scalesia*s are reduced to populations of 50 individuals or less, and are still subjected to goat grazing so their continued survival in the wild is very doubtful. Such species may be conserved *ex situ* outside the Galápagos, and cultivated or propagated by modern micropropagation techniques, so that a stock would be available for re-introduction once this is considered desirable and possible. In Table 3 some Galápagos species considered to be priority species for *ex situ* conservation measures are listed.

The *ex situ* programme is a last insurance against extinction, but at the same time as acutely threatened species are saved and made available for scientific purposes as well as for reintroduction and reforestation schemes in the Galápagos, the programme will generate a body of new knowledge, relevant and applicable to *in situ* conservation activities in the islands. Thus, the Botanic Garden of the University of Copenhagen, aims at contributing in its way to the conservation of the unique flora and vegetation of the Galápagos Islands.

Table 3. Some endemic Galápagos species of priority for *ex situ* conservation and propagation.

	IUCN Category	Main threats
Asteraceae:		
Darwiniothamnus alternifolius	V	Cattle; land use; *Psidium guajava*
Lecocarpus darwinii	E	Feral goats
L. lecocarpoides	E	Population reduced to perhaps non-viable level
L. pinnatifidus	V	*Lantana camara*
Scalesia atractyloides	E	Feral goats
S. gordilloi	V	Urban development
S. helleri subsp. *santacruziana*	E	Feral goats
S. retroflexa	E	Feral goats
S. stewartii	V	Feral goats
Cactaceae:		
Opuntia megasperma var. *megasperma*	V	Feral goats
Melastomataceae:		
Miconia robinsoniana	V	Land use; invasive plants, esp. *Cinchona succirubra* and *Psidium guajava*
Portulacaeae:		
Calandrinia galapagosa	E	Feral goats; population reduced to perhaps non-viable level
Rubiaceae:		
Psychotria angustata	E	Land use; feral mammals
Scrophulariaceae:		
Galvizia leucantha subsp. *pubescens*	E	Population reduced to perhaps non-viable level
Verbenaceae:		
Lippia salicifolia	E	Land use; feral mammals

IUCN categories according to Lawesson (1990a).

References

Adsersen, H. (1989). The rare plants of the Galápagos Islands and their conservation. *Biological Conservation*, 47, 49-77.

Anonymous, (1974). *Plan Maestro Parque Nacional Galápagos.* PNUD/FAO/ECU/71/022, pp. 1-91, Santiago, Chile.

Anonymous, (1980). *How to use the IUCN Red Data Book Categories.* IUCN, Threatened Plants Committee Secretariat, England.

Atkinson, I.A.E. (1990). Mapping and classifying vegetation in the Galápagos Islands, some considerations. In *Botanical Research and Management in Galápagos*, ed. J.E. Lawesson, O. Hamann, G. Rogers, G. Reck & H. Ochoa, pp. 101-115. Missouri Botanical Garden, St. Louis.

Bowman, R.I. (1961). *Morphological Differentiation and Adaptation in the Galápagos Finches.* Brussels, Palais des Académies.

Brockie, R.E., Loope, L.L., Usher, M.B. & Hamann, O. (1988). Biological invasions of island nature reserves. *Biological Conservation*, 44, 9-36.

Calvopina, L.H. & de Vries, T. (1975). Estructura de la población de cabras salvajes (*Capras hircus* L.) y los daños causados en la vegetación de la isla San Salvador, Galápagos. *Revista Univ. Católica, Quito,* Año 3 (8), 219-214.

Cifuentes Arias, M. (1990). Conservación y manejo de la vegetación en Galápagos. In *Botanical Research and Management in Galápagos*, ed. J.E. Lawesson, O. Hamann, G. Rogers, G. Reck & H. Ochoa, pp. 243-246. Missouri Botanical Garden, St. Louis.

Corley Smith, G.T. (1990). A brief history of the Charles Darwin Foundation for the Galápagos Islands 1959-1988. *Noticias de Galápagos*, 49, 1-36.

Eliasson, U. (1984). Native Climax Forests. In *Key Environments. Galápagos*, ed. R. Perry, pp. 101-114. Pergamon Press, Oxford.

Geist, D.J., McBirney, A.R. & Duncan, R.A. (1985). Geology of Santa Fe Island: the oldest Galápagos volcano. *Journal of Volcanology and Geothermal Research*, 26, 203-212.

Hamann, O. (1974). Contributions to the Flora and Vegetation of the Galápagos Islands. III. Five New Floristic records. *Botaniska Notiser*, 127, 309-316.

Hamann, O. (1979a). Regeneration of vegetation on Santa Fe and Pinta Islands, Galápagos, after the eradication of goats. *Biological Conservation* 15, 215-236.

Hamann, O. (1979b). On climatic conditions, vegetation types, and leaf size in the Galápagos Islands. *Biotropica*, 11 (2), 101-122.

Hamann, O. (1981). Plant communities of the Galápagos Islands. *Dansk Botanisk Arkiv*, 34 (2), 1-163.

Hamann, O. (1984). Changes and threats to the vegetation. In *Key Environments. Galápagos*, ed. R. Perry, pp. 115-131. Pergamon Press, Oxford.

Hamann, O. (1985). The El Niño influence on the Galápagos vegetation. In *El Niño en las Islas Galápagos. El Evento de 1982-1983*, ed. G. Robinson & E.M. del Pino, pp. 299-330. Quito, CDF.

Hickman, J. (1985). *The Enchanted Islands. The Galápagos Discovered.* Anthony Nelson, U.K..

Hoeck, H.N. (1984). Introduced fauna. In *Key Environments. Galápagos*, ed. R. Perry, pp. 233-245. Oxford: Pergamon Press.

Howell, J.T. (1942). Up under the equator. *Sierra Club Bulletin*, 27, 79-82.

Itow, S. (1990). Herbaceous and Ericaceous communities in the Highlands of Santa Cruz, the Galápagos Islands. In *Botanical Research and Management in Galápagos*, ed. J.E. Lawesson, O. Hamann, G. Rogers, G. Reck & H. Ochoa, pp. 47-58. Missouri Botanical Garden, St. Louis.

Krogstrup, P., Nrgaard, J.V. & Hamann, O. (1990). Micropropagation of threatened endemic and indigenous plant species from the island of Rodrigues. *Botanic Gardens Micropropagation News*, 1, 8-12.

Lawesson, J.E. (1990a). Threatened plant species and the priority plant conservation sites in the Galápagos Islands. In *Botanical Research and Management in Galápagos*, ed. J.E. Lawesson, O. Hamann, G. Rogers, G. Reck & H. Ochoa, pp. 153-167. Missouri Botanical Garden, St. Louis.

Lawesson, J.E. (1990b). Alien plants in the Galápagos, a summary. In *Botanical Research and Management in Galápagos*, ed. J.E. Lawesson, O. Hamann, G. Rogers, G. Reck & H. Ochoa, 15-20. Missouri Botanical Garden, St. Louis.

Lawesson, J.E., Adsersen, H. & Bentley, P. (1987). An updated and annotated check list of the vascular plants of the Galápagos Islands. *Reports from the Botanical Institute, University of Aarhus*, 16, 1-74.

Lawesson, J.E. & Estupinan, I. (1987). The reforestation project on San Cristóbal Island. *Noticias de Galápagos*, 45, 23.

Lawesson, J.E., Hamann, O. & Black, J. (1990). Combined recommendations from the workshop: botanical research and management in Galápagos. In *Botanical Research and Management in Galápagos*, ed. J.E. Lawesson, O. Hamann, G. Rogers, G. Reck & H. Ochoa, 291-295. Missouri Botanical Garden, St. Louis.

Lawesson, J.E. & Ortiz, L. (1990). Plantas introducidas en las Islas Galápagos. In *Botanical Research and Management in Galápagos*, ed. J.E. Lawesson, O. Hamann, G. Rogers, G. Reck & H. Ochoa, pp. 201-210. Missouri Botanical Garden, St. Louis.

Loope, L.L., Hamann, O. & Stone, C.P. (1988). Comparative conservation biology of oceanic archipelagoes. Hawaii and the Galápagos. *BioScience*, 38 (4), 272-282.

Macdonald, I.A.W., Ortiz, L., Lawesson, J.E. & Nowak, J.B. (1988). The invasion of highlands in Galápagos by the Red Quinine-Tree, *Cinchona succirubra*. *Environmental Conservation*, 15(3), 215-220.

Macdonald, I.A.W., Loope, L.L., Usher, M.B. & Hamann, O. (1989). Wildlife conservation and the invasion of nature reserves by introduced species: a global perspective. In *Biological Invasions. A Global Perpsective*, ed. J.A. Drake, H.A. Mooney, F. di Castri, R.H. Groves, F.J. Kruger, M. Rejmánek & M. Williamson, pp. 215-255. John Wiley & Sons, U.K.

Nowak, J.B., Lawesson, J.E., Adsersen, H. & de Vries, T. (1990). A two-year study of post-fire vegetation dynamics on southern Isabela, Galápagos Islands, Ecuador. In *Botanical Research and Management in Galápagos*, ed. J.E. Lawesson, O. Hamann, G. Rogers, G. Reck & H. Ochoa, pp. 123-136. Missouri Botanical Garden, St. Louis.

Ortiz, L. & Lawesson, J.E. (1987). *Informe sobre los ensayos de control de Cinchona succirubra en la parte alta de Santa Cruz.* Charles Darwin Research Station, Isla Santa Cruz, Galápagos, Ecuador (report, 10 pp).

Perry, R. (1984). The Islands and their History. In *Key Environments. Galápagos*, ed. R. Perry, pp. 1-14. Pergamon Press, Oxford.

Ramirez, G. (1988). Preliminary study of Matazarno (*Piscidia carthagenensis*) on Santa Cruz. In *Annual Report 1984-1985*, pp. 32-34. Charles Darwin Research Station, Isla Santa Cruz, Galápagos, Ecuador.

Schofield, E.K. (1989). Effects of introduced plants and animals on island vegetation: examples from the Galápagos Archipelago. *Conservation Biology*, 3(3), 227-238.

Simkin, T. (1984). Geology of Galápagos Islands. In *Key Environments. Galápagos*, ed. R. Perry, pp. 15-41. Pergamon Press, Oxford.

Utreras, M. (1987). Distribution of Guayaba (*Psidium guajava*) and application of herbicides for its control on Santa Cruz Island. *Annual Report 1983*, pp. 16-17. Charles Darwin Research Station, Isla Santa Cruz, Galápagos, Ecuador.

de Vries, T. & Calvopina, L.H. (1977). Papel de los chivos en los cambios de la vegetación de la isla San Salvador, Galápagos. *Revista Univ. Católoca, Quito*, Ano 5 (16), 145-169.

Werff, H. van der (1978). *The Vegetation of the Galápagos Islands.* Lakenman & Ochtman, Zierikzee.

Whelan, P.M. & Hamann, O. (1989). Vegetation regrowth on Isla Pinta: a success story. *Noticias de Galápagos*, 48, 11-13.

Wiggins, I.R. & Porter, D.M. (1971). *Flora of the Galápagos Islands.* Stanford University Press, Stanford.

Resumen

La flora y vegetación de las Islas Galápagos, Ecuador, se encuentran amenazadas por la utilización de ciertas prácticas de manejo de la tierra (agrícolas) y por la introducción de especies vegetales y animales foráneas. En este trabajo se discuten las medidas de conservación utilizadas en el presente para hacer frente a estas amenazas, así como la prespectiva futura de conservación de las plantas auctóctonas, para asegurar de esta manera su utilización racional.

The threatened plants of the flora of the Seychelles and their conservation

F. Friedmann

ORSTOM, Museum national d'Histoire naturelle, Laboratoire de Phanérogamie, Paris, France

The Republic of the Seychelles comprises 2 distinct phytogeographical territories: On the one hand there are the very ancient granite islands, which is partially endemic with Asian and Afro-Madagascan affinities, and on the other hand the Coralline Islands, which are of more recent geological origin, and of which the indigenous flora is a direct descendant of neighbouring regions (East Africa, the Comoro Islands, Madagascar) and with a lower level of endemism.

The forests, which formerly entirely covered the granite islands, have undergone vast changes since the settlement of the Seychelles a couple of centuries ago, and this has taken place because of the human colonization of the otherwise previously uninhabited islands. One sees very clearly in the Seychelles the well known process which one sees practically everywhere, and which accompanies population growth, that is to say land clearance, fires, the introduction of exotic species of which some like cinnamon (*Cinnamomum verum*) naturalize themselves and become ubiquitous. Only the peaks of mountains and the granite domes which are the most inaccessible, remain intact, and this represents only 5-10% of the total surface. It is above all the cultivation of the coconut palm, which has caused the almost complete disappearance of the low altitude forest, and the establishment of the cinnamon cultivation which has caused the greater part of the destruction of high altitude forests.

The Coralline Islands were similarly transformed, except Aldabra which remains relatively preserved and which today is an integral natural reserve and a site of global interest.

Without detailing here the entire history of the search for the flora of the islands, I would like nevertheless to remember a few essential dates:

TROPICAL BOTANIC GARDENS: THEIR ROLE
IN CONVERVATION AND DEVELOPMENT
ISBN 0—12—346850—7

the first botanical collections in 1840 by Auguste Perville, followed in 1871 and 1874 by John Horne, who discovered *Medusagyne oppositifolia*. In 1905 and 1908 the "Sealark" expedition, led by J.S. Gardiner, carried out important plant collections. Resident naturalists, H.P. Thomasset and P.R. Dupont, enriched the collections with several new species. The necessity for conservation measures were already being felt in 1874 when J. Horne proposed a reserve in the Vallée de Mai to protect the famous "Coco de mer", *Lodoicea maldivica*.

An evaluation of the state of the plant life was drawn up by C Jeffrey in 1961-62, with detailed proposals about placing in a reserve those plants which were the most representative of the original vegetation. These botanical collections form the basis of the national herbarium of the Seychelles.

These proposals were taken up and developed by J. Procter who also developed an important herbarium in 1970-71 and served with the Seychelles Government to establish legislation for conservation with the creation of national parks and special reserves. Between 1981 and 1987 I myself was able to contribute to research on the flora and to discover several new species.

The flora of the Coralline Islands and most especially those of Aldabra were studied in depth between 1961 and 1975 by several botanists participating in expeditions of the Royal Society, and in particular F.R. Fosberg and S. Renvoize who published a Flora of the Islands in 1980.

Extinct species

Thanks to all these contributions, the knowledge of the flora of all the Seychelles Islands is today fairly satisfactory. One notices that of 350 indigenous species of phanerogams discovered since 1840, only 5 had not been found recently (ie. between 1960 and 1987) although there are also about 80 fern species which will not be considered here, as there has been too little recent study.

- *Vernonia seychellensis* Baker (Compositae), found only by Horne in 1874 in an area that today is much altered by habitations. It has probably disappeared.

- *Lautembergia* aff. *neraudiana* (Baill.) Coode (Euphorbiaceae), was collected in 1840 and 1874 in the low altitude forest, which is today an inhabited and cultivated zone. This is apparently the only indigenous species that has disappeared before being studied and described.

- *Bakerella clavata* subsp. *sechellensis* (Bak.) S. Balle (Loranthaceae), has not been seen again since 1905, but it is a plant which is difficult to find, and it is possible that it still exists.

- *Eulophidium seychellarum* Summerh. (Orchidaceae), found on Mahé once in 1903 but never seen again in its typical form. However, on Mahé (in the same locality), on Praslin and on Silhouette, these exists an indigenous species, *E. pulchrum*, (Thou.) Summerh., which is bigger than *E. seychellarum* and may indeed be the same species.

- *Smythea lanceata* (Tul.) Summerh. (Rhamnaceae), coastal species which is distributed as widely as Fiji, and was seen for the last time in 1915. The Seychelles are at its western geographical limit. The chances of finding it are slim, because the coastline has been greatly modified.

Five species which are deemed to be extinct, (of which there is a chance of finding one or two) do not represent other than a tiny proportion of the known flora.

However, as the indigenous vegetation of the granite islands has diminished by more than 90%, some species already little known in their natural state, have today become exceptionally rare. Some could be on the road to extinction if measures are not taken to protect them. An annotated list of these species follows:

The endemic species of the granite islands (in taxonomic order)

These are only species which can be given the IUCN Red Data Book category E (Endangered) although within this category there is some variation.

Those which are under immediate threat, are listed as E (for example the *Gastonia* sp. of which only one individual is known. It is situated beside a road, growing with cinnamon plants). Others, such as *Vateriopsis* sp., have a much better chance of survival despite the fact that there are less than 50 individuals known. They are in fact distributed in several places which are not very frequented visited, and where there is a certain amount of natural regeneration, due to the long life-span of the trees, etc.

Medusagyne oppositifolia Baker (Medusagynaceae) E.

This is the only species of the family Medusagynaceae which is endemic to the Seychelles. 40 to 50 individuals have been seen on Mahé in two locations. Regeneration has only been observed once by L. Chong Seng in a place which was difficult to get to. The plants fruit regularly, and germinate in nursery conditions abundantly. However, several propagation tests carried out in the Seychelles have failed, the small plants having died after one or two years. Cultivation from seed was successful, first in the Zurich Botanic Garden, and then in the Nancy Botanic Garden (and perhaps elsewhere). The small plants do not develop unless they are placed in very humid conditions, contrary to what one might be led to believe in observing adult plants in their natural environment. After several years, the young plants become more tolerant to more normal conditions of humidity and develop well in the open (recently seen in Nancy and in Brest, where one of the bushes has flowered). It is also cultivated at Kew.

Vateriopsis seychellarum (Dyer) Yeim (Dipterocarpaceae), E.

This monospecific endemic has affinities with *Semonoporus* and *Vateria* of Asia. In the past, some of these trees grew to heights greater than 30 metres. Recently, less than 50 individuals have been seen in 5 to 6 areas on Mahé. Its natural regeneration is poor, but has nonetheless been seen in 3 places. Fruiting is abundant, and germination is immediate, but in nursery conditions the young plants die, most often after 2 to 3 years. The Forestry Service has succeeded in replanting a few young plants in the wild. The overall situation of this species is therefore a little more favourable than that of *Medysagyne*. It is also being cultivated at the botanic garden of the University of Aberdeen in Scotland.

Impatiens gordonii Baker (Balsaminaceae), E +

Only 20 to 30 individuals of this species are known in 2 places on Mahé. Happily, the cultivation of it is easy and the plant is well maintained in collections (seen recently at Brest). It is also cultivated at Kew.

Gastonia lionnetii F. Friedmann (Araliaceae), E + .

An intermediate species between *G. crassa* and *G. seychellarum*; only one adult and 2-3 young individuals are known. It is not in cultivation.

Gastonia seychellarum var. *curiosae* F. Friedmann, E + .
Gastonia seychellarum var. *contracta* F. Friedmann, E +

These two varieties are only known by the individual type of each, and neither is yet in cultivation. The three Araliaceae, of which 2 are growing a few metres from a road, can be considered as the most threatened taxa of the flora of the Seychelles.

Rothmannia annae (Wright) Keay (Rubiaceae), E.

This species has disappeared from Mahé, Praslin and Silhouette and no longer exists except on the island of Aride. The population numbers several dozen individuals but is not directly threatened because the island is a nature reserve. The Forest Service of the Seychelles has planted a few examples on Mahé. The species reproduces itself well, and can be cultivated in greenhouses fairly easily, as for example, at Nancy. It is cultivated also at Cambridge, Kew and Mascarin.

Psathura sechellarum Baker (Rubiaceae), E.
(*Psychotria sechellarum* (Baker) Summerh.).

The number of individuals known is less than 40, but there are undoubtedly more in places which have not been searched yet because they are inaccessible in the hygrophile forest of the mountain summits. It is present on Mahé and Silhouette. It is not yet in cultivation.

Psychotria silhouettae F. Friedmann (Rubiaceae), E + .

This species was discovered on Silhouette in 1985. It is distinguished from the other *Psychotria* of the Seychelles by its white fruit and its little tiny flowers. It is endemic to Silhouette. Only 3-4 individuals have been seen. It is not yet in cultivation.

Rapanea seychellarum Mez (Myrsinaceae) E.

This bush exists on Mahé and Silhouette, scattered in the high altitude forest. Less than 50 individuals have been seen. It probably exists in greater number (occurring in the same biotope as *Psathura sechellarum*).

Carissa edulis var. *sechellensis* (Baker) Pichon (Apocynaceae), E.

A small tree which does no longer exists on Silhouette, and of which the population can be estimated at less than 150. In the past the species was fairly commonly found at low altitudes, and its wood was burnt to repel mosquitos, and occasionally to heat copra drying machines. An attempt at cultivation was made at the botanic garden on Mahé. The plant grew easily without requiring a lot of care, except for one or two treatments during its young stage against cochenilles and ants.

Toxocarpus schimperianus Hemsl. (Asclepiadaceae), E.
This is a very rare species known on Mahé, Praslin and Curieuse. Two places where it has been found are contained within nature reserves. This species is not under immediate threat, although less than 50 individuals have been seen. It is not yet in cultivation.

Achyrospermum seychellarum Baker (Labiatae), E.
A little shrub known only on Silhouette in 3 or 4 places with 20-30 individuals having been seen. The species is sometimes attacked by cochenilles (*Icerya seychellatum*). It is not yet in cultivation.

Pisonia sechellarum F. Friedmann (Nyctaginaceae), E.
An endemic tree, discovered in 1983 in an enclosed valley on Silhouette. The range of this species has not been able to be precisely located, because exploration of this area is very difficult (it is very rocky). There are perhaps 100 individuals. But the fact that the species is concentrated in this area constitutes in itself a handicap for the survival of the species. The main threat, in the medium term, is the invasion of the valley by exotic plants such as cinnamon, which is already present. This plant is not yet in cultivation.

Angraecum maheense Schltr. (Orchidaceae), E.
A species described as endemic (but may only be an infraspecific taxon of *A. multiflorum*). It is known only on 2 places on Mahé and one on Silhouette. It is in cultivation in Nancy.

Hederorkis seychellensis Bosser (Orchidaceae), E.
A rare orchid found in small populations dispersed on Mahé and Silhouette. It is not yet in cultivation (an attempt at cultivation in a greenhouse has failed).

Platylepsis sechellarum S. Moore (Orchidaceae), E.
A very rare species occurring in small populations on Mahé. It could survive, at least for some time, under the cover of cinnamon shrubs.

Hypoxidia maheensis F. Friedmann (Hypoxidaceae). E.
Rare species, existing only in the hygrophile high altitude forest, in small populations, but not immediately under threat. In cultivation in Edinburgh, Brest and Nancy.

Seychellaria thomassetii Hemsl. (Triuridaceae), E.
A saprophytic species of the high altitude forest which appears sporadically. It has only been found in 1906, 1961, 1982 and 1988. It seems very rare, but is not easy to spot, and could be more widespread than is thought. Not in cultivation.

Garnotia seychellensis Hubb. and Summerh. (Gramineae), E.
A grass which grows on rocky outcrops that protrude beyond the high altitude forest, found in 4 or 5 places on Mahé and Silhouette. Cultivated in Brest.

Indigenous species (granitic islands and Aldabra).

Some species only considered to be indigenous or whose status is liable to change when they will be better known, are extremely rare either in granitic islands or on Aldabra.

I mention a tree example, described first of all as endemic from the Seychelles, under the name *Bosqueia gymnandra* Baker (Moraceae). A recent revision links this species with *Trilepisium madagascariense* DC. which probably exists West Africa as far as Madagascar and the Seychelles (and which includes 13 previously species previously regarded as distinct). In the Seychelles, the species has only recently be found and only on Silhouette (5 to 6 individuals in 3 places). In the local context one could, therefore, classify the species as category E. I would suggest that this species be included in the list of threatened species of the flora of the Seychelles, because it is an interesting element of the heritage of the flora of the island. It belongs in effect to an island population, isolated clearly from the rest of the area, and it could be that further studies will reveal certain differences.

Other species in this category:

Pseuderanthemum tunicatum (Afzel.) M. Redh. (Acanthaceae), E.
This species was found twice at the beginning of the century on Mahé and Silhouette, and it was at first species deemed to be endemic (*Eranthemum gardineri* in sched.). It was found again in Silhouette in 1982, its ecology and the location where it was found tend to indicate that it is indeed an indigenous plant which is almost identical to the African species. It exists only in one valley. It is cultivated in Nancy, where reproduction by seeds is easy.

Amaracarpus pubescens Bl. (Rubiaceae), E.
A very rare bush on Silhouette which must undoubtedly be linked with a species existing on Java. It may be an infraspecific taxon, endemic to the Seychelles. It is not yet cultivated.

Acampe rigida (J.E. Sm.) Hunt (Orchidaceae), E.
This is a very widespread species, which possibly exists from East Africa right into Asia, and has only recently been seen in one place on Silhouette. It was cultivated in the botanic garden on Mahé in 1983.

Oeoniella polystachys Schltr. (Orchidaceae), E.
A widely distributed orchid in the west of the Indian Ocean. Is only known in 3 places in the Seychelles on Mahé, Silhouette and Praslin. It is not yet cultivated.

Eulophidium pulchrum (Thou.) Summerh. (Orchidaceae), E.
See above with *E. seychellarum*). Cultivated at Nancy.

Piper sp. (Piperaceae), E.
This is a creeper which is extremely localized on Silhouette and is found in the undergrowth of the forest of *Pisonia*. Flowering has never yet been observed. Cultivation trials have failed up to now. It is without doubt indigenous and may be endemic.

Acacia aff. *pennata* (L.) Willd. (Mimosaceae) , E.
A native liane of which only 4 or 5 individuals have been seen in the undergrowth in the forest of *Pisonia*. It is in cultivation in the botanic garden of Nancy where it flowered in May 1990.

In Aldabra there are some very rare indigenous species. The taxonomic status of some species is not yet clearly established. A few listed currently as endemic must be closely related to taxa existing in East Africa, the Comoro Islands and Madagascar. Conversely, plants known under the name of *Obetia ficifolia* and considered to be indigenous, have recently been defined as the endemic species *O. aldabrensis* Friis. However, it appears that endemism in the flora of Aldabra has been overestimated. From the point of view of conservation, it seems preferable to consider all the species which are very rare on Aldabra, whatever their current taxonomic position. This is why all the species, which are more or less threatened on Aldabra, have been regrouped on a single list. Coastal or mangrove species, which are spread by marine currents, are excluded from this list.

Grewia picta Baillon (Tiliaceae), E.
(*G. aldabrensis* Baker)
About 10 individuals were recently spotted. They are often attacked by defoliating insects. It also exists on Madagascar.

Grewia glandulosa Vahl (Tiliaceae), E.
A very rare shrub on Aldabra, often attached by cochenille (*Icerya seychellarum*). It also exists on Madagascar.

Malleastrum depauperatum (Baill.) Leroy (Meliaceae), E.(*M. leroyi* Fosb.)
A rare tree on Aldabra, sometimes attacked by cochenille. It also exists on Madagascar.

Crotalaria laburnoides Klotsch (Papilionaceae), E +
It is a rare shrub that is local in distribution on Aldabra. It is threatened by wild goats. It also exists in East Africa.

Brexia madagascariensis Thou. (Brexiaceae), E +
Found once, in 1968, and not seen since. It also exists in East Africa and elsewhere.

Cassipourea lanceolata Tul. (Rhizophoraceae), E.
(*C. thomassetii* (Hemsl.) Alst.)
A small rare tree, know only to 3 localities in Aldabra. It also exists on the Comoro Islands and in Madagascar.

Eugenia capensis subsp. *aschersoniana* (Hoffm.) F. White (Myrtaceae), E + (*E. elliptica* var. *levinervis* Fos., *E. levinervis* Scott)

A rare shrub until 1975, found just once in 1983 and in 1987. Sensitive to cochenille attacks. It also exists in Zanzibar.

Peponium sublitorale Jeffrey and Page (Cucurbitaceae), E.

A very rare creeper, dispersed along the coast of southern Aldabra. It is considered to be endemic.

Psychotria pervillei Bl. (Rubiaceae), ? Ex or E +

Very rare in 1968, today it could be facing extinction following massive attacks by cochenilles. It exists in the granitic Seychelles in a slightly different form.

Salvadora angustifolia Turrill (Salvadoraceae), E.

Some individuals only on Aldabra, attacked by the cochenille. No regeneration has been observed. It is also in Madagascar.

Carissa edulis Vahl var. (Apocynaceae), E +

Only 2-3 individuals have been seen recently. Attacked by cochenille, *Icerya seychellarum*. It may be the variety *seychellensis* (Baker) Pichon which also exists in the granitic Seychelles.

Tabernaemontana coffeoides (Apocynaceae), E + (*Pandaca mauritiana* auct.)

Just one individual has recently been seen on Aldabra. It also exists in the Seychelles, in the Comoro Islands and in Madagascar.

Tylophora coriacea Marais (Asclepiadaceae), E. (*T. indica* auct.)

A few small populations near the south coast of Aldabra. It also exists in the Mascarenes.

Ehretia corymbosa DC. (Boraginaceae), E.

A few individuals occur in the east of Aldabra. It also exists in Africa, Madagascar and the Comoro Islands.

Justicia congestiflora var. *humilior* R. Ben. (Acanthaceae), E.

This has been found only on one collecting trip in a zone where wild goats are still numerous. It exists in Madagascar and its range extends into Asia.

Ruellia monanthos (Nees) Anders, (Acanthaceae), E.
Found only in 5 or 6 places. Threatened by the presence of wild goats. It also exists in Madagascar.

Pisonia aculeata L. (Nyctaginaceae), E+
One single individual was seen recently. It was under attack from the cochenille. It exists also in Madagascar and in Asia.

Bakerella clavata var. *aldabrensis* (Turrill) S. Balle (Loranthaceae)? Ex.
A variety said to be endemic of Aldabra and said locally to be common between 1968 and 1970. It has not been seen again in the few places where it was normally seen in the years from 1983 to 1987.

Viscum triflorum DC. (Viscaceae), E.
Seen in 1983 on *Pemphis acidula* in a limited area at the edge of a lagoon, attacked by the cochenille. It also exists in East Africa, in the Comoro Islands and the Seychelles.

Wielandia elegans Baillon (Euphorbiaceae) E.
A few individuals (7 to 8) were found in 1983 on the island of Esprit. It also exists in the granitic Seychelles and on Madagascar.

Obetia aldabrensis Frits (Urticaceae), E. (*O. ficifolia* auct.)
Considered to be endemic. A population of about 50 individuals has recently been spotted. It is in cultivation in Nancy.

Maillardia montana Léandri (Moraceae)? Ex or E+
It was defined as an endemic from Aldabra by Fosberg. However, a recent revision links the plants of Aldabra to a species existing in Madagascar. The only known population was devastated by a cochenille attack in 1975 and disappeared. It was not seen again in 1983 or 1987.

Acampe rigida (J.E. Sm.) Hunt (Orchidaceae), E
Known to be in 3 places, including one with a large population. It was cultivated in a botanic garden on Mahé in 1983/84.

Angraecum eburneum Bory, subsp. (Orchidaceae), E
Found on Aldabra in 3 to 4 places. It probably corresponds to a subspecies existing on the Comoro Islands. It was cultivated in the Mahé Botanic Garden in 1983.

Dioscorea bemarivensis Jum. and Perr. (Dioscoreaceae), E.
Observed in just a few places. It exists also in Madagascar. It has been cultivated from seed in Nancy.

Thus the flora of Aldabra includes at least 25 species which can be considered as threatened (regional category E) out of about 170 indigenous species. The greater part of these species also exist in neighbouring territories. However, their presence has a biogeographical significance which affords quite some importance to the natural heritage of Aldabra. In effect, certain of these species are more hygrophile than the more common species. One can consider them to be the survivors of an era when Aldabra had a more humid climate. If their rarity is a consequence of a long term climatic change, it would be regrettable if they were suddenly to disappear because of attacks of cochenille or in the jaws of wild goats.

In situ conservation

In the Seychelles the impact of human activities on indigenous vegetation has been greatly diminished to the point where it is almost acceptable by comparison with what is happening still in the other islands of the Indian Ocean. It is true that vestiges of natural vegetation only slightly degraded, are today found in very small areas and are only located in zones with very difficult access, which of course constitutes a relative amount of protection in itself.

One can still sometimes see palms (*Deckenia nobilis*) illegally chopped down for the edible "palm hearts". It is also true that the installation of television equipment on the ridge of a mountain is the cause of the destruction of several acres of primitive forest. These examples of destruction are regrettable, but perhaps they will be less frequent in future. On the other hand, a threat which remains permanent is that of exotic species, which inexorable invade the indigenous vegetation.

The first of these to become widespread was cinnamon which has already been mentioned. The areas where cinnamon does not exist are rare now. On the tracks of intact primitive forest, one often finds little plants which can take advantage of the least little bit of light coming through the vegetation canopy to develop.

Other exotics are already widely spread such as *Albizia falcataria, Alstonia macrophylla, Chrysobalanus icaco, Tabebuia pallida.* Some of these are only at the beginning of their spread and have a great future before them include *Ardisia solanacea, Memecylon caeruleum* and *Ochna kirkii.* Even an orchid such as *Dendrobium crumenatum* which colonizes the shady rocks and tree trunks at medium altitude, has come to compete with indigenous orchids by occupying their ecological niche.

These fast growing species which are efficiently dispersed by the wind or by birds, are gradually installing themselves in the primary vegetation, and given time, will end up by eliminating the indigenous species which grow more slowly and have a less robust fertility levels.

Only the manual eradication of these exotics will allow for the conservation of a few examples to the original vegetation. The material problems posed by such intervention are well known. The Conservation Service has begun to eradicate the bigger exotics such as *Albizia* in the Vallée de Mai, which is home to the most important population of *Lodoicea maldivica.* But this reserve still contains numerous exotics in places, and their eradication is a long term task.

Aldabra in the Corallian Islands, is an integral reserve and deliberate destruction can now be prevented.

When the ancient village on the island was inhabited, plant pests were introduced: *Lantana camara* which is disappearing and *Stachytarpheta jamaicensis*, which one can find near encampments, and which is clearly disseminated by the temporary occupants of camps. Long term, these exotics will have an impact on the vegetation, and one ought to be preoccupied with this from now on. A more immediate danger which threatens the survival of rare species and the integrity of the vegetation, is that of the cochenilles (*Icerya seychellarum*) introduced about 30 years ago, and the wild goats, present here over the past 100 years, and which have a tendency to reproduce in great numbers since the island became a reserve.

The Seychelles Foundation has decided to wage a war against these two menaces by launching, on the one hand a biological programme against the cochenilles, and on the other, eradication campaigns against the wild goats.

The Foundation is equally worried about the protection of rarer species threatened by cochenille attacks, which could disappear before the biological control has produced any effect. The most threatened species ought to be conserved *ex situ* for at least several decades.

Ex situ conservation

Amongst all the Seychelles species classified in the category E (endangered) those, which are the most directly threatened ought to be cultivated without any delay because in the wild they will probably only survive for just a few more years only. These species or varieties marked E are a small number.

The granitic Seychelles

Impatiens gordonii, *Gastonia lionnetii*, *Gastonia sechellarum* var. *curiosae*, *Gastonia sechellarum* var. *contracta*, *Psychotria* sp., *Acacia* sp.

One should also add, for memory, those species which are thought to be extinct (Ex):

Aldabra:

Eugenia capensis subsp. *aschersoniana*, *Carissa edulis* var., *Tabernaemontana coffeoides*, *Pisonia aculeata*, *Crotalaria laburnoides*, *Brexia madagascariensis*, *Maillardia montana*, *Psychotria pervillei* and *Bakerella clavata*.

Other species, of which the chance of survival in the near future are only a little better, if current conditions remain, ought, however, all to be cultivated and distributed to various botanic gardens.

It would, in particular, be of value and interest in the greater part of rare species of the Seychelles could be cultivated in the botanic garden in Mahé. Already several different palms and endemic *Pandanus* are already there.

In the meantime, international co-operation has already begun its work as 12 species, amongst the rarest are in cultivation in botanic gardens overseas. However, of the 15 species in the E category, only one is already in cultivation (*Impatiens*).

Conclusion

The conservation situation in the Seychelles is not unique and one finds here problems which one finds in other islands. If the pressure of human activities is not as great as it is elsewhere, the rapid spread of exotic plants and parasites is seriously compromising the long term survival of indigenous species. In the immediate future, the most threatened species are not terribly numerous (15 species) and their rescue does not represent an insurmountable undertaking. One could hope that none of them will disappear in the near future if they can be brought into cultivation.

List of species endangered in the Seychelles

Granite islands
Species apparently extinct:
Veronia seychellensis
Lautembergia aff. *neraudiana*
Bakerella clavata subsp.
Smythea lanceata
Eulophidium seychellarum

Endemic species:

Medusagyne oppositifolia, E; cult.
Vateriopsis seychellarum, E; cult.
Impatiens gordonii, E+; cult.
Gastonia lionnetii, E+
Gastonia sechellarum var.*curiosae*, E+
Gastonia sechallarum var.contracta, E+
Rothmannia annae, E; cult.
Psathura seychellarum, E
Psychotria silhouettae, E+
Rapanea seychellarum, E
Carissa edulis var., E; cult.
Toxocarpus schimperianus, E
Achyrospermum sechellarum, E
Pisonia sechellarum, E
Angraecum maheense, E; cult.
Hederorkis seychellensis, E
Platylpis sechellarum, E
Hypoxidia maheensis, E; cult.
Seychellaria thomassetii, E
Garnotia sechellensis, E; cult.

Aldabra
Native and endemic species:
Grewia picta, E
Grewia glandulosa, E
Malleastrum depauperatum, E.
Crotalaria laburnoides, E+
Brexia madagascariensis, E+
Cassipourea lanceolata, E

Eugenia capensis subsp., E+
Peponium sublitorale, E
Psychotria pervillei, E+
(to research)
Salvadora angustifolia, E
Carissa edulis subsp., E+
Tabernaemontana coffeoides, E+
Tylophora coriacea, E
Ehretia corymbosa, E
Justicia procumbens, E
Ruellia monanthos, E
Pisonia aculeata, E+
Bakerella clavata subsp., E+
(to research)
Viscum triflorum, E
Wielandia elegans, E
Obetia aldabrensis, E; cult.
Maillardia montana, E+
(to research)
Acampe rigida, E
Angraecum eburneum, E
Dioscorea bemarivensis, E; cult.

Native species (or poorly known):
Trilepisium madagascariense, E
Pseuderanthemum aff. *tunicatum*, E; cult.
Amaracarpus pubescens, subsp., E
Acampe rigida, E
Oeoniella polystachys, E
Piper sp., E
Acacia sp., E +; cult.
Eulophidium pulchrum, E

References

Baker, J.G. (1877). *Flora of Mauritius and the Seychelles*. London.

Berg, C.C. (1977). Revision of African Moraceae. *Bull. J. Bot. Nat. Belg. 47*, p. 267-407.

Fosberg, F.R. & Renvoize, S.A. (1980). The Flora of Aldabra and neighbouring islands, *Kew Bull,* Add. ser. 7, H.M.S.O. London.

Friedmann, F. (1984). Note sur les *Hypoxideae* des Seychelles. *Bull. Mus. Nat. Hist. Nat.* 4e sér. 6, p. 453-460.

Friedmann, F. (1986). Révision des *Araliaceae* des Seychelles. *Ibid*. vol. 8, p. 245-256.

Friedmann, F. (1986). Etude de la structure du périanthe des *Pisonia* paléotropicaux. *Ibid*. vol. 8, p. 383-392.

Friedmann F. (1990) Une espèce nouvelle de *Psychotria* des Seychelles. *Ibid.* vol. 12 p. 65-67.

Friis, I. (1983). A synopsis of *Obetia*. *Kew Bull.*, 38, p. 221-228.

Jeffrey, C. (1962). *The Botany of the Seychelles.* Department of Technical Cooperation, London.

Procter, J. (1974). The endemic flowering plants of the Seychelles: an annotated list. *Candollea, 29*, p. 345-387.

Summerhayes, V.S. (1931). An enumeration of the angiosperms of the Seychelles. *Trans. Linn. Soc., London*, 2nd ser. Zool. 19. p. 261-299.

Swabey, C. (1970). The endemic flora of the Seychelles and its conservation. *Biol. Conserv. 2*, p. 171-177.

Vesey-Fitzgerald, L.D.E.F. (1940). On the vegetation of the Seychelles. *J. Ecol. 28*, p. 465-483.

Botanic gardens in conservation: reintroduction into the wild

David Bramwell

Jardín Botánico Viera y Clavijo
Las Palmas de Gran Canaria, Spain

Summary

Restocking and reintroduction of individual species into their former habitats and the reconstitution of whole communities will be an important means of conserving threatened plants in the future. If it is to be considered as a valid conservation exercise, the reintroduction of species requires vigorous control, including strict guidelines and legislation. The bases of such guidelines are outlined and the requirements for thorough preliminary studies of the habitat and the elimination of the original causes of decline or loss as well as the genetic basis of reintroduction material is emphasised. Post-reintroduction monitoring and supervision needs are also considered.

Introduction

The value of the living collections held in botanic gardens for practical conservation is one of the most frequent points of contention within the plant conservation fraternity. The debate over the usefulness of the extensive collections painstakingly accumulated by curators over many years for display in gardens throughout the world seems to have concluded that their value is limited almost exclusively to education programmes. Having been involved recently in the preparation of the Botanic Gardens Conservation Strategy (Heywood, 1989) and in the review of the rôle of botanic gardens in the reintroduction of plant species into the wild, I have come to the conclusion that well curated

TROPICAL BOTANIC GARDENS: THEIR ROLE
IN CONVERVATION AND DEVELOPMENT
ISBN 0–12–346850–7

collections with accurate records can be of great value at the interface between the garden and the natural habitat, particularly as a source of material for reintroduction. This has been demonstrated on several occasions and the work carried out by the Conservatoire et Jardin Botanique de Brest on the reintroduction of *Ruizia cordata* to Réunion Island, reported elsewhere in this volume, is a good example, and many cases of plants which are either already extinct or on the verge of extinction being propagated in botanic gardens with a view to their possible reintroduction have been reported over the past few years.

Holding wild plant genetic resources in safe keeping for the future is our primary aim, and we are currently looking at the methods for doing this, garden collections, field genebanks, seedbanks etc. We are fortunate in being able to apply the technology developed by, for example, the International Board for Plant Genetic Resources, particularly for long-term storage. We must not, however, lose sight of the fact that the reintroduction to the wild and conservation of natural resources *in situ* should be our ultimate goal.

In the vast majority of cases the conservation of resources *in situ* will require interventionist management. Damage to the environment has been caused by Man and its restoration and future conservation will require Man to be involved. On a local scale, this will require us to be involved both in the field and also by providing an *ex situ* support system. It is safe to say that a considerable part of conservation activity in the 21st Century is going to be based on the concept and philosophy of restoration, repairing the damage done to the environment in the 20th Century.

Restoration biology

For the successful restoration of natural communities we will have to be able to reintroduce both plants and animals, especially higher plants and vertebrates, as these seem to be amongst the most threatened members of complex ecosystems. This opens up a completely new field of study and research, that restoration biology which is almost certainly going to be the important component of conservation biology of the next few decades. We will also need large numbers of taxonomists to classify and identify organisms. As several people have pointed out (Heywood 1971; Bramwell 1984) we are still a very long way from knowing even how many species we are dealing with, let alone having an adequate description or indications of the biology of most of those already known. I would suggest that we need what one well known modern "taxonomist" recently described to me as "bad taxonomists", those who spend their time completing inventories, describing new species from herbarium specimens, writing checklists and Floras etc., instead of spending their time in the library in search of the perfect binomial or at the computer looking for the absolute cladogram (good for the soul if not for the plants!).

From the botanic gardens' point of view the eventual use of plants and seeds held *ex situ* is a justifiable reason for maintaining them. We should, therefore, become involved in restoration biology and the restoration process in one way or another, even if only as the basic suppliers of material. We can, however, take our participation much further and become involved in the restocking and reintroduction of individual species and in attempts to restore whole communities.

The importance of reintroduction for the conservation of endangered species was recognized by European governments as far back as 1979 in the Convention on the Conservation of European Wildlife and Natural Habitats according to which (Article 11.2,a) the reintroduction of species of wild flora and fauna is to be encouraged when this would contribute to the conservation of an endangered species, provided that a study is first made to ensure that such reintroduction would be effective and acceptable. The need for reintroductions to be preceded by sound environmental studies is further emphasised by the Council of Europe who in 1985 issued a Recommendation (No. R(85) 15) on the reintroduction of wildlife species. The Recommendation was adopted by the Committee of Ministers in September 1985 and it serves as a very useful set of guidelines for the purpose of considering the botanic garden rôle in reintroduction of plants. Botanic gardens need to develop both a policy and a series of guidelines on reintroduction, restocking and restoration of vegetation and natural communities.

Reintroduction guidelines

Reintroduction is a widely used term which in practise covers at least three different situations which should be clearly separated as they involve intervention and action at different levels, they are:

1. RESTOCKING, a term used when the original wild population is depleted but has not yet become extinct, and the purpose of restocking is to increase the number of individuals in a population to a level at which it is no longer in danger from genetic depauperation, genetic or reproductive drift or other threats to chronically small populations.
2. REINTRODUCTION, applies to situations where a species is replaced into an area where it formerly occurred but has become extinct. This implies the establishment of a new population not necessarily genetically linked to the former, now extinct, one.
3. RESTORATION, is the process involving either restocking or reintroduction, or both of various elements of a plant community by means of which the former species diversity of that community is reconstructed.

From a botanic garden point of view the most important question arising from the issue of reintroduction is when and how should plants held in *ex situ* collections be used for such purposes? Because of the pitfalls involved in putting plants back into their original habitat, the need for a serious scientific justification and basis for each reintroduction or restocking is obvious and supervision and monitoring by qualified expert scientists is essential. The legal aspects of reintroduction should also be considered, especially as many of the species involved are likely to be protected by law, and any intervention in their natural habitats may require proper authorization.

The recommendations of the Council of Europe serve as guidelines for a reintroduction policy and they are outlined below.

They propose that reintroduction projects should only be carried out after conducting the appropriate ecological and socio-economic research which should:

1. determine the causes of decline/extinction
2. analyse past and present ecological characteristics
3. submit proposals, where necessary, for remedying the causes of decline/extinction

4. delimit and indicate exactly the areas in which introductions are to be carried out
5. enumerate planning, management and supervision measures to be taken before, during and after the implementation of the project
6. evaluate the possibilities of success and the possible repercussions of reintroduction
7. establish which taxa of the species should be reintroduced into a given area.

This requires a properly conceived project for each species to be reintroduced. The ecological survey which determines the original causes of decline or extinction of the species in any particular locality is fundamental to ensuring successful reintroductions. There are many causes of depletion and extinction and most of these are listed by Frankel & Soule (1981) and their list is useful, as it particularly emphasizes the importance of human impacts and of reproductive isolation in extinction

1. Biotic factors
   (a) competition
   (b) predation
   (c) parasitism and disease
2. Isolation (leading to genetic depauperation or reproductive drift)
3. Habitat alteration
   (a) slow geological change
   (b) climate
   (c) catastrophe
   (d) Man

Most causes of decline can be determined as one or a combination of more than one of the above factors. Analysis of the past and present ecological characteristics of the area concerned should indicate the extent and direction of change in habitat conditions and the possibilities of recovery. It is important to determine the exact original niche of the species and any possible modifications to that niche to ensure that reintroductions are made into the optimum habitat and not into a marginal, last refuge habitat which may not be the ideal one for the long-term survival and recovery of the species.

Following from this is the formulation of proposals for remedying the causes of decline. The correction or removal of factors such as, for example, overgrazing, spraying with insecticides which kill pollinators, trampling, excessive shade from planted trees etc. is basic to all attempts at reintroduction. The indication of the exact locality of a reintroduction might, at first glance, appear not to be of fundamental importance or even counter-productive but, in fact, constraints of ownership, permission for such activities, the taking into account of any local or national legislation on the species of area concerned etc. are all factors to be borne in mind in a project study. A further point here is that the disclosure of localities can help to prevent confusion amongst other scientists or conservationists working in the same area or on other aspects of the same species and may even help to stimulate the exchange of information, material and ideas. The possibility of an organization such as BGCS holding a register of such projects in which botanic gardens are involved should be considered as this may help to widen the sources of material and the exchange of experiences in this new and pioneering field.

The importance of planning and habitat management in modern conservation should not be underestimated and projects for reintroduction should consider them in detail. The declaration of sites of special interest, natural parks etc. can strengthen conservation efforts and where the project is to be carried out within an existing protected area then the use and management plan for the area should be taken into consideration. In such cases the local management and scientific staff should be involved to ensure co-operation with supervision and monitoring.

Though most reserves are established on the basis of what they actually contain (important ecosystems, rare species, valuable or unique landscapes etc.) with the possibilities for restoration and reintroduction a new possibility will arise, that of the creation of reserves with the specific objective of their restitution to a former, natural state or for the reintroduction into its former habitat of locally extinct species. Both these possibilities should be considered when planning local land-use etc.

There are circumstances under which the reintroduction of a species into its former habitat may have an adverse effect on other species still surviving there. The introduction of vigorous competitors or of species which might lead to the development of excessive shade are examples which come to mind and reintroductions under such circumstances should be avoided. The Council of Europe Recommendation goes as far as proposing the prohibition of any reintroduction which might have and adverse effect on the ecosystem.

When reintroductions take place, the local population and interested groups, such as environmentalists and ecology groups, should be informed and, whenever possible, involved in such projects, provided that this does not limit the possiblity for accurate scientific monitoring of the reintroduced population and control over its ultimate fate. Botanic gardens and other scientific institutions involved in such projects should, however, be prepared to provide full scientific support and supervision until such time as the reintroduced plants are integrated into the local ecosystem and there are signs of natural reproduction, such as an increase in the number of spontaneous seedlings.

One of the most important aspects of the scientific monitoring of reintroduction projects is the documentation and reporting of results. Plant reintroduction is still in an initial pioneering phase and information about failures is as important as the reporting of successes. In fact a good deal of publicity is often given to the reintroduction of the first individual to be reintroduced to the wild of species "rescued" in botanic gardens but only when populations are established and reproducing spontaneously can we claim success. Initially, more experimental reintroductions will probably end in failure than are successful and it is only by accumulating experience and knowledge that the techniques for restoration biology will be developed.

Bibliography

Bramwell, D. (1984). Biosystematics and Conservation. In *Plant Biosystematics*, ed. W.F. Grant, pp.633-641. Academic Press, Canada.

Frankel, O.H. & Soule, M.E. (1981). *Conservation and Evolution*. Cambridge University Press, Cambridge, 327 pp.

Heywood, V.H. (1971). Preservation of the European Flora: The Taxonomist's Role. *Bull. Jard. Bot. Nat. Belg.*, 41, 153-166.

Heywood, V.H. (1989). *The Botanic Gardens Conservation Strategy*. IUCN-WWF-BGCS, Gland, 60 pp.

Resumen

La mejor forma de conservar las especies vegetales en peligro de extinción en el futuro, será mediante la repoblación y reintroducción de las mismas en su habitat original, reconstruyendo así toda la comunidad vegetal.

Si el proceso es aceptado como ejercicio válido para la conservación de especies amenazadas, su reintroducción ha de ser controlada rigurosamente, y se han de subrayar las pautas y legislación necesaria para ello.

A continuación se subrayan las bases necesarias para preparar dichas pautas o directrices, así como los requisitos para la realización de un estudio preliminar a conciencia del habitat, y la eliminación de las causas iniciales de la disminución o pérdida de las especies vegetales, haciendo especial énfasis en la base genética del material reintroducido. También se considera la necesidad de vigilar y controlar el proceso después de su reintroducción.

From rescue to re-introduction; the example of Ruizia cordata

Jean Yves Lesouef

Conservatoire Botanique National de Brest
Vallon du Stang-Alar, Brest, France

Summary

The rescue of an endemic species, *Ruizia cordata* (Sterculiaceae) from Réunion island in grave danger of extinction is outlined, from collection in the wild to multiplication in cultivation at the Conservatoire Botanique National de Brest, France, and its eventual re-introduction to the wild.

Introduction

Ruizia cordata is a small tree, endemic to the island of Réunion. It belongs to the Sterculiaceae and is monotypic. It is a dioecious species of which its adaptation to dry conditions is shown by the presence of anthocyanin pigments, and the extreme dissection of the foliage in its young form.

It was discovered in the 18th century by Commerson, and at the end of the last century, Cordemoy considered it to be fairly common in the dry low altitude region. Since Cordemoy the plant had never been identified in the living state by any botanist. However, the herbarium of Rivals does have an unidentified specimen collected in 1944 at Grande Ravine.

In 1975, during a trip to the Mascarene Islands, one of my priorities was to research *Ruizia* due to its significance as a monotypic genus which was either extinct or close to it. My research was in vain. However, happily a few weeks later, Francis Friedman of ORSTOM and Harry

TROPICAL BOTANIC GARDENS: THEIR ROLE IN CONSERVATION AND DEVELOPMENT
ISBN 0–12–346850–7

Male and female specimens of *Ruizia cordata* at Brest

Gruchet of the Société Réunionnaise pour l'Etude et la Protection de l'Environnement (SREPEN) succeeded in discovering two live specimens on the Grande Chaloupe Ravine.

Subsequently, a dozen other specimens were found. With the exception of two of these, all of them have died since, most of them having had their bark peeled away or cut for uses that one can only guess about, but which one supposes to have been medical or magical. However, one should note that the area in which *Ruizia* occurred was always the victim of intense deforestation since the beginning of human occupation in Réunion. Furthermore the area has suffered considerably from either conscious or accidental introductions of plant pests or animals, such as goats, giant snails (*Achatina fulica*), *Furcraea foetida*, *Hiptage benghalensis*, etc., which pushed the natural vegetation back to the escarpments of the ravines. In the main, the regeneration of these species is either very poor or non-existent.

It seems likely that several species have totally disappeared, although tree and shrub species survive as adult specimens, sometimes very, very ancient ones. Without positive intervention, their disappearance is only a matter of time.

As far as *Ruizia* is concerned, the degradation of the environment has been such that it would be pointless to wait for spontaneous regeneration of the two surviving specimens in the wild. It is not known either whether there is a male and a female. Specific protection of these two surviving specimens is impossible, because it would be necessary to identify them, and thus they become identifiable to vandals. Only a programme of rehabilitation could save this species, a programme which would involve the intensive culture of the species, followed by reintroduction in its original environment.

In 1977 one of the specimens of the ravine of the Grande Chaloupe had cuttings taken by Yves Colette, an amateur naturalist and member of the SREPEN. Two cuttings succeeded and one was conserved in Réunion, where it was planted by Madame B. Ardon, President of the SREPEN, in the garden of a public establishment, where the plant is thriving. I brought the other plant back to Brest in 1982. A specimen discovered by T. Cadet at the Ravine des Avirons had a cutting taken by us in the same year. It has to be noted that this specimen appeared to be inaccessible on a vertical cliff that forced us to make a rope descent. However, this tree was taken away the following year, probably for reasons of black magic.

The cuttings were very difficult to root but not entirely impossible. Between 1% and 5% rooted, varying according to the clones or the times of cutting.

Two specimens were grown in the heated greenhouses at Brest. They thrived, and in 1986 and 1987 they flowered simultaneously, and we were able to see with great delight that one was male and the other female. They were pollinated by hand and produced thousands of seeds. The first germination of these, a dozen, dates back to 1986. In 1987 more than 1,400 young plants were produced at Brest, which means that the species has already been brought back from the brink of extinction.

It is worth noting that in addition to the 1,400 specimens at Brest and the female plant, planted in Réunion, two specimens have been obtained by Dominique Valck of the Conservatoire Botanique de Mascarin from seeds obtained from a tree at St. Leu that is now dead.

It should also be mentioned that from 1704 onwards the first seed germination of this species was obtained in the Jardin du Roy in Paris, and that the plant was also cultivated at the Chateau de Schönbrünn near Vienna in 1970. Nothing remains of these first attempts.

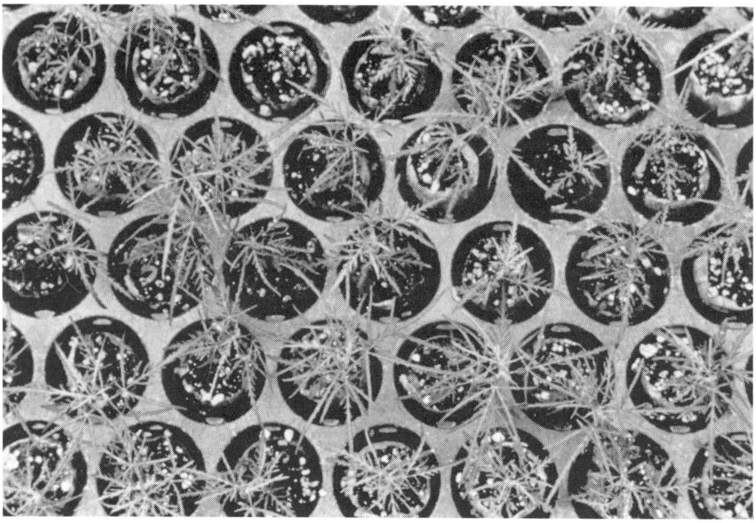

Young plants of *Ruizia cordata* at Brest

Reintroduction

With 1,400 specimens in cultivation there is the possibility of carrying reintroduction on a massive scale. *Ruizia* presents particular points when it comes to reintroduction.

1. The plant is always greatly sought after, which means that one must replanting the species in inaccessible cliffs, using ropes or, in very rare cases, in properties and localities where the plant can be assured of protection in a wild environment.
2. The plant is decorative, which allows it to be used in gardens that have a favourable climate.

In December 1988, with the help of the Ministries of the Environment and of DOM-TOM, the operation "Reintroduction" was begun. 280 plants were repatriated to Réunion. 100 were introduced on the cliffs in the dry zone, and the rest were planted in little groups in private and public gardens. The operation was undertaken by the Conservatoire Botanique National de Brest and the SREPEN.

Since then the cyclone "Firinga" has occurred and inflicted some damage on the planted *Ruizia* specimens. For this species, the damage was not as far reaching as one might have feared. This may have been due to the fact that in its young form its leaves and branches are very slender and offer little wind resistance. As well as that, its stem is very pliable.

The principal problem in the re-establishment of specimens in the wild is the competition they face from exotic species. It will undoubtedly be necessary to eradicate some of the exotic species, such as *Furcraea foetida* and *Hiptage benghalensis*, from around the areas of introduction. Experimentation will determine how often it is desirable to carry out weeding but in places with difficult access the eradication of exotic species cannot be done very frequently. After a few years the trees should be able to defend themselves.

The reintroduction work was carried out in December and the summer of 1987, and in March and April 1989. Summer in Réunion, despite its great heat, has the advantage of being the rainy season. However, in 1988-89, the summer was particularly dry, and apart from a few violent rains in December and the cyclone of January, there was not much rain. It is hoped that the planting work carried out in March and April 1989 were in less danger as they will be watered and mulched with organic matter until the summer rains, having got through the dry winter without damage. Winter in Réunion is not as hot as the summer. The little *Ruizia* plants seem well adapted to drought.

The initial population was clearly very feeble being based upon a total genepool provided by two specimens. It will be important to integrate to this genepool with the 2 plants derived from the specimen at St. Leu and the descendants of the two spontaneous survivors.

The majority of the 1,400 plants produced in Brest are due to be replanted in Réunion, for the most part in the wild, and the rest in gardens in the dry area of the island. This area is very heavily altered and marked by human occupation, with towns, gardens, many building works. The object of the operation is to initially create a population of several thousand plants throughout the area from St Denis to St Pierre both in wild and cultivated situations.

The rescue of *Ruizia cordata* can be compared to the extinction of *Astiria rosea*, another monotypic endemic genus from the Mascarenes (where it occurred in the humid upland areas of Mauritius). *A. rosea* seems to be totally extinct in the wild and has not been seen since 1860. It was cultivated in Mauritius and in England, but has not survived there either.

Amongst the other genera most endangered from the Mascarene Islands are the following:

Ramosmania heterophylla from Rodrigues surviving as one specimen in the wild and a few young plants cultivated at the Royal Botanic Gardens, Kew.

Cylindrocline spp. from Mauritius of which two species are known, one of which is extinct and the other represented by a dozen specimens.

Tetrataxis salicifolia from Mauritius, known to be in three localities and represented by less than 20 individuals.

Tectiphiala ferox, a palm from Mauritius, know to be in three localities and represented by less than 30 specimens.

Scyphochlamys revoluta from Rodrigues, known in three or four localities, and less than 30 specimens.

Nesocodon mauritianus of Mauritius, known from one site and several dozen specimens, cultivated and multiplied at the Botanic Garden of Trinity College Dublin.

All these are now in cultivation although some are still cultivated in critically low numbers.

Postscript

In April 1989, several hundred young *Ruizia* specimens were planted after the Réunion Congress, in the cliffs of the Ravine de la Grande Chaloupe and other places. At present the number of plants produced at Brest is now more than 2,000, and these are mainly destined to be reintroduced in Réunion. We will be very happy to send *Ruizia* plants to interested gardens, particularly those in the tropical and subtropical zones.

The rôle of the Fernarium as a sanctuary for the conservation of threatened and rare ferns, with particular reference to Malaysia

Aziz Bidin

Fernarium Universiti Kebangsaan Malaysia
Department of Botany
Universiti Kebangsaan Malaysia
Selangor, Malaysia

Summary

The flora of Malaysia is luxuriant and varied and contains more than 8,000 species of flowering plants, about 650 species of pteridophytes and unaccountable numbers of mosses, liverworts, lichens and algae. With the rapid development in the industrial and agricultural sectors, together with the much needed spaces for new settlements in the country, which involves clearing up of massive tracts of virgin forests, the existence of some of the plant species are constantly under threat. To alleviate the problem of extinction of certain rare and endemic species, rapid and drastic action has to be taken, such as the setting up of arboreta, botanic gardens and in case of the pteridophytes, a fernarium in the country.

The Universiti Kebangsaan Malaysia Fernarium (c.32 acres) which was recently established in the University's Forest Reserve, encompasses a variety of habitats. To date about 150 species of pteridophytes have been included collected from all over the country, as well as some foreign pteridophytes. While the emphasis has been on the germplasm collection of rare and threatened pteridophyte species, the Fernarium will also serve as an outdoor laboratory for morphological, cytological and phytochemical investigations, with a view towards developing a thorough and satisfactory classification for the Malaysian pteridophytes.

TROPICAL BOTANIC GARDENS: THEIR ROLE IN CONVERVATION AND DEVELOPMENT
ISBN 0—12—346850—7

Peninsular Malaysia's vanishing forests

The tropical rainforest is luxuriant, intricate, and varied. It is one of the most complex ecosystems in the world. In Malaysia alone there are more than 8,000 species of flowering plants, about 650 species of pteridophytes and unknown numbers of mosses, liverworts, lichens and algae. This rainforest occupies ca. 61% of the total land surface area of Malaysia (Chin, 1988), of it, only a small proportion is protected and listed as a national park. To date, Peninsular Malaysia has only one national park, i.e. Taman Negara (National Park) which is about 4,343 sq. km. or 3% of its land area. This figure is relatively low when compared to the neighbouring countries, 8-9% in Thailand and 6% in Indonesia (Anon. 1986). Recently the Malaysian Government has agreed to declare the 92,000 hectare Endau-Rompin Forest as a national park. The area is of utmost importance from the conservation point of view since it sustains the largest breeding population of Sumatran rhinoceros and large tracts of undisturbed forest which contains many Bornean elements, not to be found in forests elsewhere in Peninsular Malaysia.

Elsewhere (as is the case in many developing countries), millions of hectares of primary forest has been felled and cleared for agricultural, industrial and settlement purposes. The limestone hills and marble deposits in the north of Peninsular Malaysia, covering an area of about 260 sq.km. is also partly being quarried for the extraction of road building materials and other purposes. The floristic composition of these limestone hills and marbles varies from one site to another and with mining and quarrying activities the flora is in real danger in terms of area depletion and extinction due to changes in their special edaphic conditions requirements.

The primary forests in the various groups of islands around Peninsular Malaysia (Langkawi, Pangkor, Tioman, and Perhentian) are also being threatened with the rapid developments of tourism industry, new roads, hotels and townships are sprouting at the expense of the forests.

In depth scientific research and studies of the Malaysian tropical rainforest mainly centred on the higher plant group with immense economic importance. Relatively little is known of the lower plant groups, especially the pteridophytes (ferns and fern-allies). As regards the ferns, their study have been of recent date and it may be said to begin in 1954 with the publication of the first edition of R.E. Holttum's book

"Flora of Malaya, II. Ferns" in which he enumerated and discussed about 500 species of ferns found on the mainland of Peninsular Malaysia. Since then the number of systematic studies of local ferns have not been encouraging, and the sporadic accounts that appear from time to time are either too general and unsystematic in approach or touch on specific taxa or geographical areas only (e.g. Holttum, 1959, 1963, 1978, 1980, 1981, and 1985; Parris, 1986; Bidin, 1984a, 1984b, 1984c and 1985a and 1985b; Bidin, et al. 1988; Piggot, 1988). The ferns on the numerous islands around the country's shorelines were barely touched in Holttum's book, and to date only the ferns of Carey Island, Langkawi group of islands and Aur (in the Tioman group of islands) were intensively surveyed (Bidin & Jaman, 1983, 1986, 1988; Bidin, 1987).

With ceaseless human activities, as exemplified by the development of Pulau Langkawi as a tourist resort on the west coast of Peninsular Malaysia as well as at other various places on the mainland, certain rare and endemic species are now known to face the problem of extinction in Peninsular Malaysia (Table 1).

Table 1: Some of the threatened ferns and fern-allies in Malaysia

Locality Species	Langkawi Islands	Tioman Islands	P. Malaysia Main Range	East Malaysia
1. *Lycopodium carinatum* L.	-	-	National Park (Pahang)	-
2. *Botrychium daucifolium* Wall. ex Hook.	-	- Highlands	Gunung Jasar Cameron	-
3. *Platycerium holttumii* D. Jonch. & Hennipm.	Langkawi	-	-	-
4. *P. wallichii* Hook.	Langkawi	-	-	-
5. *P. platylobum* A. Bidin & R. Jaman	Langkawi	-	-	-
6. *Drynari bonii* Chr.	Langgun	-	-	-

Table 1: Some of the threatened ferns and fern-allies in Malaysia (cont.)

Locality Species	Langkawi Islands	Tioman Islands	P. Malaysia Main Range	East Malaysia
7. *Pseudodrynaria coronans* (Wall.) Ching	Gunung Raya	-	Gunungi Jera	-
8 *Bolbitis malaccensis* (C. Chr.) Ching	Gunung Raya	Gunung Kajang	-	-
9. *Tectaria variolosa* (Hook.) Copel.	Timun Langkawi	-	-	-
10. *Quercifilix zeylanica* (Houtt.) Copel.	-	Tioman Aur	-	-
11. *Adiantum lamrianum* A. Bidin & R. Jaman	-	-	-	Trus Madi Sabah
12. *Cyathea oosora* Holtt.	-	Gunung Kajang	-	-
13. *Angiopteris angustifolia* Presl.	-	-	Maxwell Hill Fraser's Hill	-
14. *Ophioglossum simplex* Ridl.	-	-	Panti	-
15. *Christensenia aesculifolia* (Bl.) Maxon	-	-	Maxwell Hill	Mulu Sarawak
16. *Tectaria inopinata* Holtt.	-	-	- Sabah	Keningau
17. *Schizaea wagneri* L.	-	-	Gunung Jerai	-

Conservation measures

However vast the richness of flora and fauna of any country, without proper management of its land use, the disappearance of many plant and animal species is inevitable. As for the ferns and fern-allies, to alleviate the problem of extinction of the rare and endangered species, at least three conservation measures must be considered.

1. Preservation of existing habitats
2. Collection and storage of spores for future uses
3. Collection of germplasm in a managed garden (as a living collection).

Researchers and plant scientists have to determine what rare and endangered species that are known to occur in restricted zones. Their habitats that are being threatened must be documented and adequate steps taken for their preservation. New forest reserves must be listed by the government and the existing reserves should be upgraded to national parks.

At the Universiti Kebangsaan Malaysia, the collection of fern germplasm was started in 1980 with the setting up of the University glasshouse complex. However, due to limited space and conditions the plants did not do well in enclosed and artificial conditions and the setting up of a managed fern garden in as close to a natural environment as possible was thought necessary. The idea was realised in 1988 and, with the permission gained from the university authorities to use part of the Universiti Kebangsaan Forest Reserve for the purpose, a Fernarium was established.

The concept

The Fernarium of Universiti Kebangsaan Malaysia, Bangi has been set up to conserve the rare and endangered ferns and fern-allies in Malaysia and also to include the great diversity, or at least representatives, of Malaysian and tropical fern flora into a managed garden, and to present it in a manner that is scientific, educational and aesthetic.

Scientific programme

The introduction of the fern and fern-allies into the Fernarium will mostly involve plants from the wild that have never before been cultivated. It will present a great opportunity to undertake research in the sowing, growing and propagation of such plants, especially rare and threatened species. Opportunities to develop horticultural types through hybridization and selection will be emphasised. Besides this, the Fernarium will also serve as an outdoor laboratory for morphological, anatomical, cytological and phytochemical investigations with a view to developing a thorough and satisfactory classification for the tropical ferns and fern-allies in general and for Malaysia in particular.

Educational programme

The ferns and fern-allies introduced into the Fernarium are located according to sections, and in each section the plants are arranged systematically by family. Each accession is provided with both its scientific name and its vernacular name. Common names are given for the benefit of visitors not familiar with botanical terms. Whenever possible the origin, uses and its status are also mentioned. In this way it is hoped that the visitors can appreciate the enormous variety of the tropical ferns.

Aesthetic value

As in any other botanic garden, the essence of the garden is that it must be beautiful. To achieve that the Fernarium has been landscaped in order to create a variety of scenic features - lakes, streams, walking tracks, meadows and terraces. These various features will also provide the different ecological bases needed for species with varied habitat requirements.

The project

The Fernarium, covering an area of about 22 acres, is situated just inside the University main campus, adjacent to the Faculty of Life Sciences building. It is part of the Bangi Forest Reserve assigned to the University for teaching and research activities.

The existing vegetation in the Fernarium, as well as in other parts of the Forest Reserve, has been described as a "logged over lowland dipterocarp forest simulating its climax formation". Among the dominant emergents are *Shorea leprosula*, *S. macroptera*, *Dipterocarpus crinitus*, *Alstonia scholaris* and *Dyera costulata*. Patches of *Eugissona tristis* are found scattered in the area due to the early logging activities. Secondary fringe vegetation which consists of *Dicranopteris linearis*, *Macaranga gigantea* and *Trema orientale* predominates the sides of the Fernarium.

The understorey within the Fernarium consist of members of Leguminosae, Annonaceae, Rubiaceae, Palmae, Euphorbiaceae, Dilleniaceae and many others. The ground layer of herbaceous plants consists of mainly members of Araceae, Vitaceae and pteridophytes. Epiphytic pteridophytes are common. Two natural lakes are found in the Fernarium and served by two streams coming from Bukit Rupa - the highest point (c. 100m) in the Forest Reserve.

The original fern flora of the Fernarium

This tiny stretch of forest supports a relatively high number of ferns and fern-allies. The ground flora consists of a number of terrestrial pteridophytes, notably selaginellas and the thelypteroids such as *Pronephrium*, *Sphaerostephanos*, *Mesophlebion* and *Christella*. Among the Dennstaedtiaceae, *Lindsaea*, *Pteris*, *Asplenium* and *Adiantum* are well represented. The common tree ferns of the tropics is represented by *Cyathea alternans*. This fern can be easily recognised by their massive, erect trunk and fine feathery palm-like fronds. It commonly grows in gullies and on well-shaded slopes. The small, erect coral-like fern *Schizaea* is well-distributed on the forest floor, especially along trails.

One of the most striking features amongst the original ferns in the Fernarium is the abundance of epiphytic ferns. The most common is the nest fern, *Asplenium nidus*, and the stag's horn fern, *Platycerium coronarium*. Both of these are nestled or attached onto the trunks. The basal ascending fronds are in the form of a basket or a nest and are held on by a massive short stem to the host tree. Dead leaves falling from the host tree are caught in the basket of foliage. These dead leaves rot slowly to form a rich decaying mass to provide nutrients to the roots of the ferns. The small, delicate and beautiful *Antrophyum* are abundant on the slender branches of trees partially exposed to sunlight through breaks in the forest canopy. Besides these, the forest is also rich in filmy ferns such as *Trichomanes* and *Hymenophyllum*.

The heliophytic and the early colonizer ferns flourish on the forest fringes which are directly exposured to the sum. Here, the most abundant and conspicuous are the well-known *Dicranopteris linearis*, *Lygodium microphyllum*, *Blechnum orientale*, and *Pityrogramma calomelanos*. *D. linearis* produces dense thickets on steep slopes, *L. microphyllum* scrambles over other plants and through the grasses, whilst *B. orientale* and *P. calomelanos* occur singly or in patches or bare ground.

Cultivation and propagation

The Fernarium is divided into three main sections. The first section which comprises approximately 5 acres is reserved for economic ferns. These are the edible ferns, ferns with medicinal properties and the vine ferns - those where the rachises are used in the making of a variety of handicrafts. The edible ferns are represented by four species, i.e. *Stenochlaena palustris*, *Diplazium esculentum*, *Ceratopteris thalictroides* and *Acrostichum aureum*. In these species the young fronds are eaten, after boiling, as vegetables.

To date, at least 76 species belonging to 44 genera and 13 fern families in Malaysia are known to have medicinal properties, and are used either alone or as an ingredient in a medicinal mixture containing a varieties of herbs (Bidin, 1985b, Burkill, 1935). These ferns are gradually being introduced into the Fernarium, among others are *Drynaria sparsisora*, *Asplenium nidus*, *Helminthostachys zeylanica*, *Platycerium coronarium*, *Cibotium barometz*, *Schizaea dichotoma*, *Angiopteris evecta* and *A. angustifolia* (Fig. 1).

Peelings from the rachises of the vine fern, *Lygodium*, are widely used in the making of handicrafts such as handbags, purses, trays and strings, among the rural people in Malaysia.

The second section comprises the nursery and collections of the rare and threatened ferns and fern-allies. It covers an area of about 10 acres. The nursery is equipped with 20 prothalli chambers for sowing spores and sections for propagation. The plants introduced into this section are collected from various localities in the country where their existence is presently threatened or where they are known to be threatened in the near future due to the opening of the forest. Fo far most of the accessions have been gahered from Langkawi Island, Cameron Highlands and

Fig. 1. *Angiopteris angustifolia*, a medicin al fern planted at the entrance of the Fernarium

Fraser's Hill. These are areas where massive physical developments are taking place for the tourism industry. Among the ferns and fern-allies already introduced here are: *Platycerium coronarium*, *P. wallichii*, *P. holttumii*, *Photinopteris speciosa*, *Osmunda vachellii*, *O. javanica*, *Angiopteris angustifolia*, *Lycopodium phlegmaria* and *Brainea insignis*.

Ferns and fern-allies are indisputably the most beautiful of all non-flowering plants, especially from the aesthetic appeal of their fronds. The third section of the Fernarium, (c. 7 acres) is reserved for ornamental and introduced ferns. Amongst the many species gradually being introduced into the Fernarium are many species and varieties of Maidenhair ferns (*Adiantum*), *Pteris*, *Microsorium*, *Asplenium*, *Cyathea*, *Nephrolepis* and *Selaginella* (Fig.2). With regard to foreign ferns and

Fig. 2. A variety of ornamental ferns in the Fernarium

fern-allies, viable spores have been obtained from various localities in the world through the Royal Botanic Gardens, Kew. The original as well as the introduced ferns and fern-allies of the Fernarium are listed in table 2.

Table 2 List of ferns and fern-allies grown in the Fernarium, Universiti Kebangsaan Malaysia

Psilotaceae
Psilotum nudum L.
Psilotum complanatum Sw.

Selaginellaceae
Selaginella padangensis Hieron
Selaginella intermedia (Bl.) Spr.
Selaginella willdenowii (Desv.) Baker.
Selaginella stipulata (Bl.) Spr.
Selaginella roxburghii (Hk. & Grev.) Spr.

Table 2 List of ferns and fern-allies grown in the Fernarium, Universiti Kebangsaan Malaysia (cont.)

Lycopodiaceae
Lycopodium nummularifolium Blume
Lycopodium phleqmaria L.
Lycopodium squarrosum Forster
Lycopodium phyllanthum Hook. & Arn.
Lycopodium clavatum L.
Lycopodium cernuum L.

Ophioglossaceae
Helminthostachys zeylanica (L.) Hook.
Botrychyum daucifolium Wall. ex Hook. & Grev.
Ophioglossum pendulum L.

Marattiaceae
Angiopteris evecta (Forster) Hoffm.
Angiopteris angustifolia Presl

Osmundaceae
Osmunda javanica Blume
Osmunda vachellii Hooker

Hymenophyllaceae
Hymenophyllum serrulatum (Presl) C. Chr.
Trichomanes cf. singaporeana (v.d.B.) v.A.v.R.
Trichomames javanicum Bl.

Schizaeaceae
Schizaea dichotoma (L.) Sw.
Schizaea digitata Sw.
Schizaea wagneri Selling
Lygodium polystachyum Wall. ex Moore
Lygodium circinnatum (Burm. f.) Sw.
Lygodium longifolium (Willd.) Sw.
Lygodium borneense v.A.v.R.
Lygodium cf. *merrillii* Copeland
Lygodium microphyllum (Cav., R. Br.
Lygodium flexuosum (L.) Sw.

Dicksoniaceae
Cibotium barometz (L.) J. Sm.

Cyatheaceae
Cyathea contaminans (Wall.) Copel.

Table 2 List of ferns and fern-allies grown in the Fernarium, Universiti Kebangsaan Malaysia (cont.)

Cyathea alternans Wall.
Cyathea recommutata Copel.
Cyathea latebrosa (Wall. ex Hook.) Copel.
Cyathea moluccana R. Br.

Dipteridaceae
Dipteris conjugata Reinw.

Gleicheniaceae
Dicranopteris curanii Copel
Dicranopteris linearis (Burm. f.) Underw. var. *subpectinata*
Dicranopteris linearis (Burm.f.) Underw. var. *linearis*

Cheiropleuriaceae
Cheiropleuria bicuspis (Bl.) Pr.

Polypodiaceae
Aglaomorpha heraclea (Kze) Copel.
Colysis acuminata (Bak.) Holttum
Selliguea lima (v.A.v.R.) Holttum
Crypsinus enervis (Cav.) Lopel.
Drynaria sparsisora (Desv.) Moore
Drynaria rigidula (Sw.) Bedd.
Goniophlebium percussum (Cav.) Wagner & Grether
Goniophlebium prainii Bedd.
Photinopteris speciosa (Bl.) Presl
Lecanopteris carnosa (Reinw.) Bl. var. *pumilla* (Bl.) v.A.v.R.
Microsorium punctatum (L.) Copel.
Microsorium musifolium (Bl.) Ching
Microsorium congregatifolium (v.A.v.R.) Holttum
Microsorium superficiale (Bl.) Ching
Platycerium holttumii De Jonch. & Hennipm.
Platycerium andinum Baker
Platycerium coronarium (Koenig) Desv.
Platycerium wallichii Hook.
Pyrrosia nummularifolia (Sw.) Ching
Pyrrosia penangiana (Hook.) Holttum
Pyrrosia stigmosa (Sw.) Ching
Paragramma longifolia (Bl.) Moore
Phymatosorus nigrescens (Bl.) Pichi-Serm.
Phymatosorus longissima (Bl.) Pichi-Serm.
Thylacopteris papillosa (Bl.) Kunze

Table 2 List of ferns and fern-allies grown in the Fernarium, Universiti Kebangsaan Malaysia (cont.)

Loxogrammaceae
- *Loxogramme avenia* (Bl.) C. Presl

Adiantaceae
- *Adiantum caudatum* L.
- *Adiantum capillus-veneris* L.
- *Adiantum philippense* L.
- *Adiantum latifolium* Lam.
- *Adiantum trapeziforme* L.
- *Adiantum malesianum* L.
- *Taenitis dimorpha* Holttum
- *Taenitis blechnoides* (Willd.) Sw.
- *Antrophyum callifolium* Blume
- *Syngramma alismifolia* (Pr.) J. Sm.
- *Pityrogramma calomelanos* (L.) Link.
- *Ceratopteris thalictroides* (L.) Brongn.
- *Vittaria scolopendrina* (Bory) Thwaites

Aspleniaceae
- *Asplenium phyllitidis* D. Don
- *Asplenium salignum* Bl.
- *Asplenium batuense* v.A.v.R.
- *Asplenium polydon* Forster
- *Asplenium longissimum* Bl.
- *Asplenium tenerum* Forst.
- *Asplenium scortechinii* Bedd.
- *Asplenium* cf. *caudatum*
- *Asplenium viviparum*
- *Asplenium nidus* L.

Athyriaceae
- *Diplazium esculentum* (Retz.) Sw.
- *Diplazium asperum* Blume
- *Diplazium cordifolia* Bl.
- *Diplazium accedens* Bl.
- *Diplazium bantamense* Bl.
- *Diplazium tomentosum* Bl.
- *Diplazium crenatoserratum* (Bl.) Moore
- *Diplazium malaccense* Presl.

Blechnaceae
- *Blechnum finlaysonianum* Hook. & Grev.
- *Blechnum indicum* Burm.

Table 2 List of ferns and fern-allies grown in the Fernarium, Universiti Kebangsaan Malaysia (cont.)

Blechnum vestitum (Bl.) Kuhn
Blechnum fraseri (A. Cunn.) Luerssen var. *philippinense* Christ
Brainea insignis (Hk.) J. Sm.
Stenochlaena palustris (Burm. f.) Bedd.

Davalliaceae
Davallia divaricata Bl.
Davallia trichomanoides Bl.
Humata repen (L. f.) Diels
Humata heterophylla (J. Sm.) Desv

Dennstaedtiaceae
Histiopteris incisa (Thbg.) J. Sm.
Microlepia speluncae (L.) Moore var. *villosissima* C.Chr.
Pteridium aquilinum (L.) Kuhn

Lindsaeaceae
Lindsaea heterophylla Dryand

Lomariopsidaceae
Lomagramma sumatrana v.A.v.R.
Bolbitis heteroclita (C. Press) Ching

Elaphoglossaceae
Elaphoglossum callifolium (Bl.) Moore

Oleandraceae
Nephrolepis acutifolia (Desv.) Christ
Nephrolepis davalloides (Sw.) Kze
Nephrolepis falcata (Cav.) C. Chr.
Nephrolepis cordifolia (L.) C. Presl
Nephrolepis excaltata
Oleandra pistillaris (Sw.) C. Chr.

Pteridaceae
Acrostichum aureum L.
Pteris ensiformis Burm.
Pteris ensiformis Burm. cv.
Pteris semipinnata L.
Pteris biaurita L.
Pteris venulosa Blume
Pteris holttumii C. Chr.

Table 2 List of ferns and fern-allies grown in the Fernarium, Universiti Kebangsaan Malaysia (cont.)

Tectariaceae
- *Ctenitis vilis* (Kunze) Ching
- *Tectaria devexa* (Kunze ex Mett.) Copel.
- *Tectaria griffithii* (Bak.) C. Chr.
- *Tectaria singaporeana* (Wall. ex Hook. & Grev.) Copel.
- *Pteridrys australis* Ching
- *Pleocnemia irregularis* (C. Presl) Holttum
- *Pleocnemia olivaceae* (Copel.) Holttum
- *Quercifilix zeylanica* (Houtt.) Copel.

Thelypteridaceae
- *Christella subpubescens* (Bl.) Holttum
- *Pronephrium triphyllum* (Sm.) Holtt.
- *Pronephrium repandum* (Fee) Holtt.
- *Mesophlebion trichopodum* (C. Chr.) Holtt.
- *Mesophlebion chlamydophorum* (C. Chr.) Holtt.
- *Christella arida* (D. Don) Holtt.
- *Christella papilio* (Hope) Holtt.

Grammitidaceae
- *Ctenopteris alata* (Bl.) Holttum
- *Ctenopteris* sp.

Salviniaceae
- *Salvinia cucullata* Roxb.

Azollaceae
- *Azolla pinnata* R. Br.

Equisetaceae
- *Equisetum ebile* L.

Propagation techniques

Ferns and fern-allies are propagated in several ways: by means of spores, division, buds and plantlets. For large scale production, propagation by means of spores is used. In harvesting spores, it is important to keep the sample pure and free from contamination. This is done by washing the frond with mature spores under running water then transferring it into packets made of grease-proof paper before air drying it. The spores shed in the packet then are transferred into the sowing pot and quickly covered by a watch glass. Spores of many species remain viable for a long time if kept dry at room temperature. For instance, spores of *Adiantum* and *Asplenium* are known to be long-lived. Soil used for sowing (two parts peat, one part loam and one part river sand) is steam sterilised to kill the spores of fungi, algae and bryophytes, to eliminate competition for the germinating prothalli. Watering is by means of capillary action - the pots are placed in a trough of shallow water. The cultures are kept in prothalli chambers. Prothalli produced (the time taken varies according to species) are left to self-fertilised and produce new progeny or else kept for hybridization.

Vegetative propagation by means of division, buds and plantlets is more common, especially for plants which do not normally produce fertile spores. Stolons, runners or rhizomes from the bases of older plants can be easily cut and readily produce new roots and form new plants. Other ferns, such as the mother ferns, *Adiantum caudatum, A. rhizophorum, A. incisum, A. malesianum* and *Asplenium viviparum*, bear little bulbils or plantlets on the surface of their fronds. These young plants may form roots on contact with soil.

Conclusion

The main priority in the establishment of thc Fernarium is the conservation of ferns and fern-allies, be it the domesticated forms, i.e. the ornamentals or the more important wild species attributed with medicinal properties. Conservation of the rare and endangered ferns and fern-allies is emphasized.

References

Anon. (1986). Editorial. *Malayan Naturalist*, 37(1), 1.

Bidin, A. (1984a). Ferns of family Osmundaceae in Peninsular Malaysia. *Nature Malaysiana*, 9(1), 28-31.

Bidin, A. (1984b). A chromosome count for *Osmunda* from Peninsular Malaysia. *Fern Gazette*, 12(6), 360-362.

Bidin, A. (1984c). The importance of leaf forms in the subdivision of the genus *Adiantum* L. *Sains Malaysiana*, 13(3), 279-289.

Bidin, A. (1985a). The fern of the genus *Brainea* in Peninsular Malaysia. *Nature Malaysiana*, 10(2), 26-28.

Bidin, A. (1985b). *Paku-pakis Ubatan Semenanjung Malaysia*. 67 pp. BDP, Kuala Lumpur.

Bidin, A. (1987). A preliminary survey of the fern flora of Langkawi Islands. *Gard. Bull. Sing.*, 40(2), 77-102.

Bidin, A. & R. Jaman. (1983). Notes on the ferns and fern-allies of Carey Island. *Malayan Naturalists*, 37(1), 13-19.

Bidin, A. & R. Jaman. (1986). A new species of *Platycerium* from Peninsular Malaysia. *Gard. Bull. Sing.*, 39(1), 149-151.

Bidin, A. & R. Jaman (1988). The coastal ferns of Pulau Aur, Johor. *Nature Malaysiana*, 13(3), 22-25.

Bidin, A., R. Jaman & M.K. Salleh (1988). A new species of *Adiantum* from Trus Madi Range, Sabah. *Gard. Bull. Sing.*, 41(2), 45- 48.

Burkill, I.H. (1935). *A Dictionary of Economic Products of the Malay Peninsular*, vol. 1 & 2, 2444 pp. Crown Agents for the Colonies, London.

Chin, S.C. (1988). Looking after our forests. *The Star Newspaper, Kuala Lumpur*, Dec. 28, 17.

Holttum, R.E. (1954). *Flora of Malaya II, Ferns*. Government Printing House, Singapore.

Holttum, R.E. (1959). Gleicheniaceae and Schizaeaceae. *Fl. Malesiana*, II, 1, 1-16.

Holttum, R.E. (1963). Cyatheaceae. *Fl. Malesiana*, II, 1, 65-176.

Holttum, R.E. (1978). Lomariopsis Group. *Fl. Malesiana*, II, 1, 255-330.

Holttum, R.E. (1980). The fern family Thelypteridaceae in Malaya. *Gard. Bull. Sing.*, 33, 1-30.

Holttum, R.E. (1981). The fern genus *Tectaria* in Malaya. *Gard. Bull. Sing.*, 34(1), 132-147.

Holttum, R.E. (1985). Two new species of *Tectaria* from limestone in Peninsular Malaysia. *Gard. Bull. Sing.*, 38(2), 145-148.

Parris, B.S. (1986). Grammitidaceae of Peninsular Malaysia and Singapore. *Kew Bulletin*, 41(3), 491-517.

Piggot. A. (1988). *Ferns of Malaysia in Colour*, 458 pp. Tropical Press, Kuala Lumpur.

Resumen

La flora de Malasia es muy rica y variada, y contiene más de 8.000 especies de fanerógamas, cerca de 650 especies de pteridófitas, y un número incalculable de musgos, hepáticas, líquenes y algas. Algunas especies vegetales se encuentran en constante peligro de extinción, debido al rápido desarrollo de los sectores agrícolas e industriales, y la destrucción masiva de bosque virgen como consecuencia de la presión demográfica. Se necesita tomar una decisión drástica y rápida, como puede ser la creación de arboretums, jardines botánicos, y en el caso de los pteridófitos, la creación en el país de un “pteridario”, para aliviar el problema de la extición de ciertas especies raras o endémicas.

El “pteridario” de la Universidad Malaya de Kebangsaan (con cerca de 13 hecrateas) fue recientemente fundado dentro de la Reserva Forestal de la Universidad, y presenta una serie de habitats variados. Hasta la fecha se han introducido en el “pteridario” 150 especies de pteridófitas recogidas por todo el país, así como especies foráneas. Aunque el énfasis del “pteridario” es la colección de germoplasma de especies de pteridófitos raras y amenazadas, éste servirá como laboratorio de campo para las investigaciones de caracter morfológico, citológico y fitoquímico, con el objetivo de desarrollar una clasificación majorada de los pteridófitos malayos.

The Lipizauga Botanical Sanctuary: a conservation project in Papua New Guinea

Revd. Canon N. Cruttwell

Director, Lipizauga Botanical Sanctuary
Eastern Highlands Provincial Government
Goroka, Papua New Guinea

Introduction

The Lipizauga Botanical Sanctuary (L.B.S.) is a type of botanic garden within a nature reserve, the Mt Gahavisuka Provincial Park near Goroka in the Eastern Highlands Province of Papua New Guinea (P.N.G.). Its purpose is to preserve the highland flora of P.N.G. as a form of genebank. The flora of Papua New Guinea has been estimated as 90% endemic, and the highland flora is probably even more so but it is also far less well known than the lowland flora. The Lipizauga Botanical Sanctuary is a unique project for P.N.G. Whereas there are botanic gardens in Port Moresby and Lae on the coast, these are not particularly conservation-oriented. The only other projects in the highlands are the Highland Orchid Collection at Laiagam and the Kaiap Orchid Lodge. These are much smaller and concentrate more on the sale than on the scientific study of orchids, to which family they are mainly restricted.

Situation

The L.B.S. is situated on a spur of Mt Otto just outside Goroka town, and reached by a 4-wheel-drive road, soon to be upgraded. The altitude of the Sanctuary is approximately 2,000 to 2,600 m. so the climate is sub-tropical. The rainfall is about 2,800 mm. and the temperature cool to cold at night (to 10°C). The soil is clay over volcanic rock with a pH

TROPICAL BOTANIC GARDENS: THEIR ROLE IN CONVERVATION AND DEVELOPMENT
ISBN 0—12—346850—7

below 5.00. It is ideal for growing Ericaceae, of which many occur naturally.

Figure 1. Lipizauga Botanical Sanctuary (part of Mt Gahavisuka Provincial Park) car park with the author, and Mid Mountain Rainforest behind.

Vegetation

The natural vegetation is savannah grassland on the lower ridge and mid montane rainforest above, dominated by *Castanopsis acuminatissima* and *Lithocarpus* sp. It is extremely rich, including many unidentified species. The trees are covered with epiphytes, including orchids and rhododendrons. The scenery is very beautiful with the Asaro Valley beneath and vistas of mountains all around, including Mt. Wilhelm, the highest in P.N.G. (over 4,400 m).

Development

The Park was opened in 1983, with the road still incomplete. We started planting accessions to the Botanical Sanctuary as early as 1981, but had

Figure 2. Lipizauga Botanical Sanctuary visitor centre with garden staff

not decided on its boundaries or its name at that time. We regarded the Park and Sanctuary as one unit. Later on we decided to define an area for planting *ex situ* specimens and to use the separate name. The total area of the Park is 77.4 hectares, but we hope to enlarge it to about 150 ha. in the future, to include a small area of upper montane forest. The L.B.S. is about 5 hectares and fenced to keep out pigs, which do a lot of damage. The central visitor area is shared by the Park and the L.B.S. We have a joint Visitor Centre and two staff residences, for both. Within the L.B.S. we have two plant houses, wired and shaded but without glass.

Accessions

We have introduced about 3,000 accessions in about 500 taxa, many of them single plants or cuttings. Not all of these have survived. We do not yet have adequate facilities for propagation. We specialize in orchids and rhododendrons. We try to grow a few specimens of each orchid in the plant houses, and plant the rest outside. Rhododendrons (of which all but two are endemic) are mostly planted outside, except for a few

Figure 3. The Orchid House

small epiphytes. Orchids include many species of *Dendrobium* and *Bulbophyllum*, with many other genera. Many species are undescribed. We also grow begonias, impatiens, gingers, etc. and we try to grow some of the alpine species from over 3,000 m., but not always successfully.

Endangered species

At this stage in the development of this country, few species can be regarded as endangered, but many could soon become so, especially those which are rare, local or attractive to collectors. We are making a point of collecting such species and growing them at the Sanctuary.

Staff

The staff consists of a director, a curator and one trained forestry assistant.

Records

The Director has kept records of all taxa growing *in situ* and all accessions in notebooks and a card index and has collected many herbarium specimens, which have all been sent to the National Herbarium, Lae, and other centres. He has also taken thousands of photographs of P.N.G. flora as well as making drawings of some plants of particular interest. He is now engaged in computerizing all records, which will be included in the BGCS database of botanic garden conservation plant holdings, incorporated by means of the International Transfer Format for Botanic Garden Records. This work is being carried out on an Amstrad PC IBM compatible computer which was donated by the British High Commission. A set of P.N.G. postage stamps designed by the Director was issued in 1990, depicting four of the rhododendron species grown in the Sanctuary.

The future

We are a long way from fulfilling the aims of the Sanctuary, even though we have not done so badly in a mere six years. We had hoped to start sealing the road to the Sanctuary in 1989, but this has not yet been done (1990) owing to P.N.G.'s current financial problems. We are constructing a small lake, but are having problems with leakage. We need some sort of a camping or guest house, as many visitors wish to spend the night to see the wildlife, notably including Birds of Paradise. We need a proper irrigated propagation facility for seeds and cuttings. If we are to propagate orchids we need a small laboratory with electric power or gas. A mini-hydro is being installed, but will only give enough power for fluorescent lighting. It would also be convenient for visiting scientists, of whom we get many, to prepare and examine specimens and for other forms of research. My hope is that one day it might become a botanical or ecological research centre on similar lines to the Wau Ecological Institute but with a highland botanical emphasis. There is no such centre in a truly highland area (i.e. above 1500 m). However, we cannot do these things at present with our limited staff and a shoestring budget. The Provincial Government alone has not the resources to fully develop it. It needs national government support, as well as overseas involvement and sponsorship. It is now internationally recognized, and we have many overseas visitors.

However, the project is making steady progress, and is, we believe, and important one from a scientific point of view. Already a number of new species have been described from the Sanctuary, and more are being named each year, especially in the genus *Bulbophyllum*. In this time when an increasing amount of rainforest is being clear-felled and large areas cleared and polluted for mining, such conservation reserves are important for the preservation of species. It must not be allowed to lapse, or degenerate into a mere ornamental park. It needs to be continued on a scientific basis, as a valuable conservation project.

Part 4

Plant Trade and CITES

CITES: Plant trade and botanic gardens

Ger Van Vliet*

Hortus Botanicus
Botanic Garden of the University
Leiden, The Netherlands

Ever since mankind began creating and using gardens for his pleasure there has been a continuous search for the most interesting plant species to place in it. Even after the invention of glasshouses (with orangeries as the most primitive form) and the consequent increased interest in tropical plants, only a limited amount of people could afford this luxury. The number of specialized traders, who collected the oddities from the wild, was equally low. With the hazards of transport, and so on those few plants which survived the trip were tenderly cared for and used for artificial propagation. The greatest interest was in succulents, which could easily survive the boat journeys and many of the larger trading firms in Europe have a long tradition stemming from this period. This century (particularly since the Second World War) saw an enormous and quick change in this pattern. Central heating and the economic prosperity of many were the main cause for an increasing demand for exotics in the developed countries. Not only Cactaceae, but also orchids, the carnivorous plants and cycads were much sought after. The number of specialized, and most unfortunately also non-specialized, traders continues to increase, and they all go to the source countries, collect their material or buy it cheap on the local markets (fill a number of suitcases) and return to sell for a quick profit. A market grows. Those who buy the plants are often only interested in the more common species for a short period only, or not at all. Then they want the rarities, not yet present in the collection of their neighbours. There is a trade in large quantities of all kinds of species.

---

* Present address: Plants Officer CITES Secretariat P.O. Box 78 CH 1000 LAUSANNE 9, Switzerland.

TROPICAL BOTANIC GARDENS: THEIR ROLE IN CONVERVATION AND DEVELOPMENT
ISBN 0—12—346850—7

Rare species, often command high prices, and what is more relevant for this discussion, quick and considerable profits for the trader. The very large columnar or barrel cacti are available in Europe for US$ 2,000 to 4,000, and single plants of *Paphiopedilum sanderianum* were recently offered for US$ 3,500. All were guaranteed to be wild. The prospect of an easy profit (often untraceable for taxes) brings again and again irresponsible traders and middlemen to the market. In many of the source countries there is a reasonably sound structure to acquire orchids or cacti for trade. The local people collect the plants (orchids, cacti or bulbs, depending on the country) and these are brought via middlemen to the international trader. This may mean a modest addition to their meagre income. But it gets out of hand when specialized collectors force local people to reveal the localities of rare orchid species in the Philippines, for example, and bring (smuggle) these to Thailand and Hong Kong. Or when Chinese and Burmese species are deliberately smuggled to Thailand and Japan. Not to mention the many stories about cacti collectors and the way in which full truck loads apparently seem to enter the USA unnoticed.

Nature is facing a wide variety of threats, of which trade is only one. The tropical forests are quickly disappearing. We all are very aware of that. The traditional method of shifting cultivation is perhaps the least dangerous one. In South America conversion of the forests for ranching or for electricity schemes are more serious. In these cases, however no plant trade is involved. The modern logging operations in Asia and Africa clear large areas of forest, but often less that 25% of the felled wood is of sufficient quality to be traded; the rest is left to rot or is burned. Many are inclined to think that these valuable hardwoods are used at least for more or less durable purposes like flooring, furniture etc. but only recently it became clear that Japan (a mass consumer of tropical wood) uses it for single-use purposes like crates to transport shoes - often to the very countries they took the wood from, or to make casings for concrete work.

This useless trade in rare woods is a trade in species or genera. CITES can play a rôle here. Forbidding trade in the wood of such species could help to save the rest of the forests where they are standing. It is shocking to see what it destroyed and it is therefore not surprising that in certain regions the local people are more and more often opposing this abuse of their natural resources. And it also, with reason, gives support to the oft-heard statements: "If you see how all orchids/cacti are destroyed what difference does it make if I take a few" or "Let us take all those plants out and grow them ourselves". Ultimately this may lead to some other more positive activities. The first salvage operation for orchids has been started in Mexico, although I feel that in this particular case commercial aspects play an equally important rôle.

The rôle CITES can play to help in the protection of what is left to us is a clear one. I cannot and will not confront you with all the aspects of plant trade. In November 1988 the Plants Committee of CITES met in Kew to discuss these issues. The meeting lasted four days and we did not even cover all the ground we had to. Since then the 7th Meeting of the Conference of the Parties was held in Lausanne (October 1989). During this meeting the Plants Committee held its second meeting. The third is scheduled for July 1990 in Venezuela. Here I can only briefly describe how CITES works and touch on some of the problems. CITES, the Convention on International Trade in Endangered Species of Wild Fauna and Flora, is a trade convention, and it assists in controlling (the species on Appendix II) or restricting (the species on Appendix I) the trade in endangered species. (See Wynstekers, 1988 (new edition, 1990) or Favre, 1989). Which species will be on which Appendix is decided by the countries who, voluntarily, have become a Party to CITES. Persons making the decisions at the biennial CITES Conference of the Parties to the Convention are official representatives of their Government, subject to political instructions. A ratification of CITES by a country means that all those countries must build into their legislation the instruments necessary for the proper implementation of CITES. Basically, the appendices were, for the greater part, drafted during the first meeting in Washington, in 1973. There are still species on it which do not really belong there, which are irrelevant for trade, or which can not be effectively controlled because you cannot recognize them in the form they are traded in, or because they are not rare at all. This matter is being looked into by the Plants Committee and some necessary amendments have already been accepted by the meeting in Lausanne. One point needs to be particularly stressed. A CITES listing does not mean that trade in that plant is forbidden. Only if a species is listed on Appendix I is commercial trade prohibited. Even in this case exemptions, specifically intended for the creation of a parent stock for artificial propagation, can and will be made. CITES is designed to control at the international borders the movement of threatened species for trade, but also by private collectors. In the case of cacti and succulents I am disappointed to hear only the intentions, not the explicit will of all members of the International Organization of Succulents to stop the removal of rare species from the wild. Many botanic gardens and private persons alike apparently still want these in their collections.

I have already mentioned that Orchids and Cacti (those on Appendix I excepted) are listed as a group on Appendix II. There is a varying degree of threat for many of these species. Because of the look-a-like problem and to facilitate control, all are listed on Appendix II. Trade in species on Appendix II is simply regulated and is still possible. A country of origin issues a document in which they state that they have no objection to the export of a given set of indigenous wild plant specimens by a given person. This document must cover that shipment. With this document import into another country is possible. Customs should inspect and check whether the data mentioned in the export certificate conform to the contents of the shipment. Some countries (like those of the EC) also require an import permit, issued by the country of import. That is meant as an extra control. However, the factual control element is always the export document in relation to the contents of that shipment. But there are some countries like Brazil, Colombia and also Australia who **explicitly forbid the export of all wildlife**, and wild plants fall under this category. The reasons for such a ban are many and different. It may be based on the idealistic view of protecting their wildlife, but may as well be the consequence of serious problems caused by excessive smuggling of cat or crocodile skins. There is perhaps a strange contradiction when we consider the export ban by Brazil and the destruction of the Amazonian forests. But any country has a legal right to impose such a ban. And we must respect their decisions and help them carry it out. It cannot be stressed enough that the philosophy of CITES has embedded in its structure the requirement to help any source country in the implementation of such a ban. Still there are people who think they are above the law. They still want to trade in wild plants. So they obtain documents stating that the wild plants they want to bring out of that country are artificially propagated. The document hence no longer covers the shipment. Now, it is the case that CITES certainly does not want to hinder the trade in artificially propagated plants. So there is a lenient attitude towards these shipments. Consequently, shipments of wild plants often enter uninspected under false documentation as artificially propagated, not only from Asia and South America but also from Australia, USA and countries of Europe. When such a shipment is inspected, the counterfeit is observed and the import is not allowed; the shipment is held by Customs and confiscated. Sometimes the importer is very well aware of these falsifications, sometimes he acts in good faith and is surprised that the shipment is not in accordance with regulations. In normal business life the shipment should be returned, because of inadequate delivery. But the orchid trade is different, apparently because the greater part is prepaid. The buyer insists on importing, thus making his own actions illegitimate. The knowledge needed to determine if a plant is wild collected or not (in the form it is traded), is or should be available in the botanic gardens and associated institutes. I am aware that many never realized that such information could be relevant. But others, with great experience in the field, know these differences very well. Unfortunately, their contacts with traders are often such that they are not willing to assist Customs. Apparently many collections are still dependent on the import and trade of wild collected plants. A proper identification handbook is needed. But this should be different from the traditional Identification Manual CITES has developed for identification of animals. Plants are traded in a form in which they can not easily be identified, i.e. without flowers. The United Kingdom has recently decided to provide £5,000 for the production of a booklet 'Plants in trade, a guide to CITES'. We expect this to appear in 1991.

Any confiscation leads to three points:

1. A given amount of plants to be kept until it has been established in the courts if the importer is guilty or not. Finding proper housing creates a serious problem if an active enforcement programme is carried out. Confiscated shipments may vary from a few plants only, to 4 tightly packed suitcases with orchids, to a 60 foot container filled with Cycads.

2. The second problem is the frustrated trader, telling all kinds of tales to try and prove his innocence. And there is always someone around somewhere who can tell why CITES does not work or should not even exist for plants. They forget that the very source of their frustration and anger is the fact that CITES did work

properly. The true and pretended ignorance and the swamp of tell-tale stories has created many of the present misunderstandings about CITES, not only among the amateurs, but also in the more professional circles of taxonomists and botanic gardens. One of those is that CITES is disadvantageous to the people in the countries of origin, who lose a chance to make some money. In cases where trade is forbidden (because the species is on Appendix I or because of the fact that a country has an export ban) this is true where wild plants are concerned. In other instances trade can go on, but in a regulated way! This regulation only excludes illegal trade. It furthermore provides the mechanism by which trade in artificially propagated plants can be stimulated. For the source countries it is very relevant that this artificially propagation of plant species is further developed. The vast amount of hybrids, produced in the tropics, indicates that know-how and facilities are adequate. It would be unwise to allow nurseries in the northern hemisphere to dominate this market.

3. Finally there is the legal end of the process. Ignorance about the plant trade is considerable as I have already mentioned. Often it is therefore necessary to convince the country of origin that something suspicious has happened. Many do not want to admit that. If I may take an Australian case as an example: 25 indigenous tropical orchid species, totalling 1,500 large plants, all wild, travelled under a special licensed label as artificially propagated. First the Australian Government reacted with a telex to the trader. "The Dutch scientist involved did not know how things operated and these plants were grown outside under semi-natural conditions. They only looked wild". I wondered how one can grow tropical orchids outside in regions with frost. So I phoned them, talked to the responsible person, who admitted he was at loss because he was not familiar with the plant trade. It was suggested to him to send a specialized botanist from a botanic garden, I believe from the Australian National Botanic Gardens, Canberra. His report confirmed that the exporter did not propagate these species artificially and had violated Australian law. The actions by the Netherlands scientist were justified. But it is still uncertain whether Australia

is going to prosecute this exporter. This example shows that not only must some efforts be made, but it is justified to do so.

I often get the impression that many with an opinion about CITES actually do not know how many wild plants are really traded and with what effects. Stephanie Paine (1989) gave the right circumscription: "Plants are the Cinderellas of Conservation". But the real Cinderella story turned out to have a happy ending! Is anyone aware how many wild Paphiopedilums (originating from everywhere in Asia) are exported yearly from Thailand, how many are imported by Japan, the Netherlands or the USA, and how many die in the process of collection and transportation to the markets? Do we know how many bulbs of *Galanthus* or tubers of *Cyclamen* are harvested yearly in Turkey; and with what consequences for the natural vegetation? Do we know how many tons of wild collected succulents were shipped yearly to Germany from Madagascar? Do we know how many container loads of bromeliads travel yearly from South America to Europe? The great majority of these plants are certainly not collected from locations where the original vegetation was to be destroyed for one purpose or another. They are collected from their natural habitats for the sole purpose of trade. Many must have had the experience of finding an orchid species (wild) in the collection of a trader, knowing that it occurs in a nature reserve only and that it should not be in trade at all. And there are some other questions the botanic gardens should ask themselves. Countries which are party to CITES must have a Scientific Authority, which advises the government on all aspect of nature conservation related to trade, a botanist should be a member. But if there is a Scientific Authority, is there a botanist in it? And if so, is that person just a professor in plant taxonomy or does he know something about the plant trade and nature conservation itself? I can but will not answer that question for you. You find out yourself. And if you know the answers, we can come back to the first query: does CITES work for plants? The questions about the Scientific Authorities are of course general in nature but are, in particular, aimed at the main importing countries: Japan, Europe, and the USA. In these countries there is a vast intellectual knowledge assembled in and outside the research institutes. And we should use that to help the source countries. We must appreciate the position of these source countries, understand how they work, and accept the limitations. There is great concern in South America, Asia and Africa for the conservation of their natural heritage. Let there be no doubt about that. There is a contradiction between a total ban on the export of wildlife or of a particular species, and the continuing destruction of the rain forests, true. And we can spend hours, days talking about that. But what we actually **should** do is help those and other countries who have a rule that none of their wildlife may be exported for trade, to implement that protective measure - via CITES or any other mechanism (although I know of no other one that could work so adequately).

CITES exists; presently 106 countries are Party to this convention. Contrary to so many other schemes developed to conserve our nature, it is **not** a pattern of thoughts awaiting expansion and action. The international network is ready, and its structure is such that everyone can easily participate. If every country would simply co-operate we would see a few direct and important results:

- International trade in wild plants would be reduced to an acceptable level not detrimental to the natural population;
- the local people would still have the possibility of making some money collecting for the trade;
- the production of artificially propagated material would increase.

What botanic gardens can, no **should** do, is simple:

- Convince your country to implement CITES properly for plants and animals.
- Insist that you are at least involved with the Scientific Authority in your country (and if you have to hurt the ego of a professor in botany with only a remote idea about the plant trade, so be it);
- if your country is not yet a Party, start the discussion about accession;
- assist in the nursing of confiscated plants, because if you and your country are successful with the implementation of the Convention there will be seizures;
- make sure that you produce seed from these confiscated plants and seek reliable growers and traders who will produce the seedlings to flood the market with the needed artificially propagated material;

- think about your own behaviour and give an example to the amateur world;
- don't buy wild plants unless they are really needed. This will mean that you must also reconsider your collection policy;
- stimulate amateur societies to use the word 'collection' in the most positive sense – not to collect or buy wild plants;
- co-operate with each other and the CITES Secretariat to unveil the patterns of illegal trade. In doing so you may loose a few contacts you thought were friends. Don't worry. The rare plants will benefit. We should not fool ourselves. We all know that there is no real need to continue the collecting of wild plants. Most, if not all can be grown artificially and will thrive much better than those wild-collected.

There is no harm in wild collecting **if it is done** in such a way that the natural population is maintained. We must also be rational, and not expect any person to starve for the maintenance of a part of nature. Trade brings them needed income. But it would certainly be a better solution if artificial propagation of all species is established in the countries of origin. We know how much expertise is available if we look at the vast amounts of hybrids produced in Thailand, the Philippines and India. If we see how easily Cactaceae can be grown in the Canary Islands, then why not in Brazil and Mexico as well? Stopping the wild trade where it is detrimental to the original wild population is the best way to create this switch to artificial propagation. I have pointed my finger more or less accusingly to the botanic gardens in general. I did this on purpose and in some cases with good reason. We, as botanic gardens, should take our responsibilities seriously and make CITES work, reduce the trade in wild plants to an acceptable level, and make sure our countries adequately implement CITES. The structure is there, waiting. All you have to do is take some action. The Netherlands form a main link in the trade of artificially propagated plants and is also one of the main importers of wild-collected plants. They have taken their responsibility seriously where CITES is concerned. Other countries are challenged to do the same. And I would love to come back to this subject at the next conference, – or preferably earlier – to discuss how well CITES *does* work for plants.

References

Favre, D.S. (1989). *International Trade in Endangered Species*. Martinus Nyhoff, Dordrecht.

Paine, S. (1989). The Case of the Stolen Slippers. *New Scientist*. 24 June, 48-53.

Wynstekers, W. (1988). *The Evolution of CITES*. CITES Secretariat (new edition August 1990), Lausanne.

TRAFFIC-France against the traffic of tropical plant species

C. Sastre and F. Moutou

TRAFFIC-France, c/o WWF-France
Paris, France

Summary

TRAFFIC-France, sponsored by WWF-France and FRANCE-NATURE-ENVIRONNEMENT strives to fight illegal trafficking of animal and plant species on the French national territory. Many small offences and contraventions of CITES regulations are caused by ignorance of the legislation, so every endeavour is made to provide information for education and through the media to increase awareness of the extent and problems caused by illegal trade in endangered species. TRAFFIC-France collaborating with the French government administration (Environmental Ministry, customs, etc.), various nature protection associations and other TRAFFIC organizations, particularly in Europe. Among plants, orchids and succulents are the illegal trafficker's favourites. It is suggested that botanic gardens and conservatories set an example for others to follow in their respect for the provisions of the CITES Convention.

Following the excellent chapter by Van Vliet on the CITES agreements, in this paper we will discuss another aspect of the Convention: putting it into effect. As in the paper on the actions of WWF-France in the West Indies, we will discuss some of the activities undertaken by those who implement the recommendations in France, suggested by various organizations active in the field of conservation, such as IUCN and WWF-International.

TROPICAL BOTANIC GARDENS: THEIR ROLE IN CONSERVATION AND DEVELOPMENT
ISBN 0–12–346850–7

In the framework of activities by WWF and TRAFFIC in the French national territory, we are attempting to apply the recommendations they have laid down, always bearing in mind that French law has to be observed, and considering also those activities recommended by the Secretary of State for the Environment, where these are in agreement with the principles of WWF.

Amongst the causes of the disappearance of wild species, one of the most destructive is the collection of wild plants, especially their collection for commercial reasons.

As soon as the CITES Convention has been signed and acceded to by a member country it must be applied. In France, we have had to design laws for the application of CITES, and this was done in 1978. The enforcement of these laws was placed in the hands of an appropriate government department. After this, information has to be made available to the public and to users of wild species, such as zoological and botanical gardens.

Aware that the passing on of information between different links in a chain is not always easy, and conscious of the continuing commercial trade of wild threatened species, despite the recommendations of CITES and the various laws drawn up in different countries, a proposal was made to create a network to fight all aspects of such illegal traffic in endangered species. This international network was established 45 years ago between IUCN and WWF and is called TRAFFIC.

TRAFFIC-France was born out of an initiative by WWF-France and France-Nature-Environnement (FNE) (formerly the Fédération Française des Sociétés de Protection de la Nature - FFSPN). It is funded by WWF and has its head office with WWF-France itself, at Versailles.

The director of WWF-France was for 2 years J.B. Dumont, and he has now been replaced by F. Moutou. C. Sastre is the botanist on the scientific committee of WWF-France and also on the administrative committee of the office of TRAFFIC-France, thus assuring that botanical matters are given the attention they deserve and to ensure adequate communication between the different scientific disciplines.

What are the means of action of TRAFFIC-France?

As soon as the new organization TRAFFIC-France was established, WWF-France and FNE (which was then still FFSPN) published a pamphlet for public distribution announcing the establishment of the office of TRAFFIC-France. The pamphlet contained the information that 3,200 animal species and 40,000 plant species are threatened with extinction, that the international trade in wildlife is the second largest cause of disappearance of this wildlife, and that collectors and tourists also contribute to this loss. We believe that information has a very important rôle to play in the implementation of CITES as many small offences are committed, simply through ignorance of the law and of international conventions.

This is the reason why much effort is put into making information available to the media and we had the satisfaction of seeing the contents of our pamphlet reproduced by different newspapers and journals. Even a few radio and television programmes have been devoted to our emergence as an organization, particularly in the light of two specific exploits, one being the seizure of monkeys and parrots at the Roissy-Charles de Gaulle airport, which provoked the arrival, in the middle of the night, of the then Minister of the Environment, who came to the airport at our request. After a few days in the menagerie of the Muséum National d'Histoire Naturelle in Paris, these animals were sent back to their countries of origin.

We feed the press also with communiqués on issues such as the illegal trade of wild collected orchids. About a year ago there was a European Orchid Congress. It included some high level scientific communications as well as stands from which orchids were sold. A year prior to that, in 1987, there was another orchid congress held in Nancy at which we learned, after the event, that *Paphiopedilum* orchids of Vietnamese origin had been sold at a Thai stand with CITES licences, produced simply at people's request. Thus, during the Paris congress, and with the agreement of the State Secretariat for the Environment, customs officers carried out inspections in the various sales stands. They found various contraventions of the Washington Convention (the Washington Convention is a name by which CITES is often known), and in consequence issued various summonses.

We are willing collaborators with the various specialist naturalist associations, and amongst these orchid enthusiasts are particularly active, such as:

- The Société Française d'Orchidophilie, with which WWF-France has edited a brochure entitled "S.O.S. threatened plants" calling for the provisions of the Washington Convention to be respected.
- Assocation Guadeloupéene d'Orchidophilie (A.G.O., l'Abeille d'Or, named after the local name of *Oncidium* orchids) which has collaborated on the production of a regional list of orchids for Guadeloupe. This list has now appeared, and a legal wrangle is going on to try to stop a network of orchid traffickers who sell plants gathered from the wild without CITES licences, near the airport of Raizet.
- Association des Orchidophiles Martiniquais (club Ionopsis, orchids are represented in Martinique by two wonderful species). During an exhibition which will be held in February of 1990, we are planning a WWF-TRAFFIC stand.
- The Société Guayanaise d'Orchidophilie, which regularly informs us about its activities.

Documents have been produced with other TRAFFIC bureaux such as the one produced with TRAFFIC-Belgium and entitled "Si vous aimez la Nature comme moi, elle n'en a plus pour longtemps" (If you like Nature like I do, it won't be here for much longer).

Finally, we are using WWF-France channels to promote TRAFFIC-France and a special edition of "Panda", the WWF-France newsletter has been produced on the theme "Traffic of nature, trade of death" and given out to participants at congresses. It contains several articles, in particular one concerning the dangerous passion that exists for orchids. The central part of the newsletter is devoted to the Washington Convention. It was produced as a supplement of which several thousand copies were printed and widely distributed.

We are also collaborating with the customs service in two ways:

- by arranging demonstration sessions, showing species which feature on CITES lists (zoological training sessions or workshops were given with great success in 1988, and a botanical one is planned for 1989),

- by acting as experts to offer advice. The following two examples illustrate how such advice is often necessary:

a) We uncovered a case concerning *Saussurea lappa*, an Indian species featuring on Appendix 1 of the Washington Convention. With the aim of cheating the authorities and getting around the law the exporter featured *Saussurea lappa* in his catalogue and marked on his parcels the obsolete name for this species, the name of *Costus*, that is to say the present day name of an important genus of Zingiberaceae.

b) We noted the arrival in France of a consignment of plants sent from Madagascar enclosing species which feature on CITES lists, imported without licence and said to have been cultivated in gardens. They were in fact orchids, *Pachypodium* (Apocynaceae) and Euphorbiaceae belonging to species which are difficult to reproduce in cultivation.

Finally, the office of TRAFFIC-France is involved in activities concerning the creation of dossiers on various subjects, which it then submits to appropriate administrations, for example: the trade in frog legs, tortoises and their products, ivory and traffic or trade in various overseas French departments for France (Guyana, Guadeloupe, Polynesia).

Conclusions

These few examples show how in France we are trying to apply the CITES agreements.

It is important that botanic gardens, both public and private should set an example for the rest of the world by respecting CITES regulations while continuing in their work on the introduction of threatened plant species.

We must also be vigilant at the local level, for example in Réunion itself. In the main Réunion market one can see tree fern trunks sold as flower pots. These ferns belong to species, not yet exactly under threat, but they are endemic to the Mascarene Islands, and the limited number of populations are constantly being reduced in size and number. Although the commercial exploitation of these fern trunks is only happening at a local level and the damage is therefore not too great, if the trade were to become international, there would be a very real risk of seeing these species become rare or even disappear. The same could be said for other local species, such as orchids, for example, which are always in great demand.

Resumé

TRAFFIC-FRANCE, coparrainé par WWF-FRANCE et France-Nature-Environnement, s'efforce de combattre les trafics illégaux d'espèces animales et végétales sur le territoire national. Beaucoup de petits délits existent par ignorance des réglements, aussi l'effort est porté sur l'information en utilisant tous les moyens médiatiques possibles. TRAFFIC-FRANCE collabore avec diverses Administrations (Douanes, Secrétariat d'Etat à l'Environnement), diverses Associations de Protection de la Nature et les autres Bureaux TRAFFIC. Orchidées et succulentes sont les favorites des trafiquants. Il est proposé que les Jardins botaniques et Conservatoires montrent l'exemple en respectant les lois et la Convention CITES.

Resumen

TRAFFIC-FRANCE co-patrocinado por el WWF-FRANCE y France-Nature-Environnement trata de combatir los trafico en especies de plantas y animales ilegal en el territorio nacional. Muchos delitos se cometen por ignorancia de la legislación reglementos, entonces se hace un esfuerzo para proveer informacion usando de todos medios possibles. TRAFFIC-FRANCE colabora con varias administraciones (aduanas, Secrétariat d'Etat à l'Environnement), varias Sociedades de Proteccion de la Naturaleza y con otras oficinas de TRAFFIC. Dentro de las plantas, las Orquidaceas y succulents son las más frecuentes en el trafico en plantas prohibidas. Se propone que los jardines botánicos y los conservatorios den un ejemplo en respectar la ligislación y la Convencion CITES.

Part 5

Botanic Gardens and Orchid Conservation

Orchids in botanic gardens – necessity or luxury?

Joyce Stewart

Sainsbury Orchid Fellow
Royal Botanic Gardens
Kew, Surrey, U.K.

Summary

An introductory paper about the family Orchidaceae and the value and importance of orchid collections in Botanic Gardens is presented. The rôles such a collection can play in the varied work of botanic gardens are discussed. The importance of botanic gardens for the practice of *ex situ* conservation techniques with orchids in the face of current regulations are emphasised.

It is particularly appropriate that orchids and the rôle they play in botanic gardens have been chosen as a special theme for this Congress. Apart from being the largest family of flowering plants, the Orchidaceae is also an extremely interesting one, about which not nearly enough is known. On the island of Réunion more than 100 species have been recorded, many of them endemic, but no detailed account of them has yet been written. More study is needed of the orchids of Réunion and in the adjacent islands before the account of the family for the Flora of the Mascareignes can be completed. In addition to the indigenous plants, many tropical orchids from other countries are grown as garden plants in Réunion. There is obviously great interest in these beautiful flowers in our host country and it is a pleasure to see such variety in cultivation in gardens large and small.

Orchids have sometimes been described as the Royal Family of plants. But the Orchidaceae is more than a special family, warmly regarded by its devotees; it is also a very large one. One can only make a guesstimate

TROPICAL BOTANIC GARDENS: THEIR ROLE
IN CONVERVATION AND DEVELOPMENT
ISBN 0—12—346850—7

of the total number of species in the family, because lists are still so incomplete for so many countries. Twenty five thousand is the figure usually quoted, some ten per cent of the total number of all flowering plants. In the last few years, more than one hundred new species have been described every year. At the same time. revisionary work has reduced some names to synonymy. How many species are dying out before they become known? One must add that more than 80,000 man-made hybrids have now been registered through the Royal Horticultural Society. These hybrids provide the bulk of the cultivated orchids of amateur and commercial cultivation today. It is only the botanic gardens and a few dedicated amateurs, who are still interested in growing species collected from the wild.

My presentation begins with a brief overview of the great diversity that exists within the orchid family. At the same time, it should become apparent that orchids can be a great asset in a botanic garden and a considerable responsibility.

Like many Europeans, my interest in orchids began with the wild species near my home, the terrestrial species or *Orchis*, *Ophrys* and *Dactylorhiza* that flower in the early summer. Some fifty species are known in the British Isles but many of them have become rare in recent years and ten are on the endangered list. Nearly thirty years ago, I went to live in Africa, where I found a great variety of terrestrial orchids in many different habitats. They can be found on the sandy beaches of tropical shores at the Equator, near the snow line at the tops of the highest mountains, and at many places in between. A few, from the Cape, were known to the younger Linnaeus. Others are still being discovered and described today.

It was in Kenya that I first encountered epiphytic orchids and began to appreciate the enormous diversity that exists among these tropical plants. Orchids are the largest group of epiphytic plants. More than half the family, at least 14,000 species, are entirely dependent on woody vegetation for their niche in the natural ecosystem. Habitat destruction is thus the greatest threat to their continued existence. This Congress should consider whether there are ways in which botanic gardens can be involved in the rescue or salvage of wild orchid plants, in areas where destruction and "development" are taking place?

Orchids vary greatly in the sites they occupy within the habitat, some seeking light positions within the canopy while others flourish in the shade. They are all perennial plants, many of them rather slow-growing. They vary greatly in size. the largest known is *Grammatophyllum speciosum* whose leafy stems are often 10m. long. It grows exuberantly in the Singapore Botanic Garden but in the wild the sheer weight of the orchid plant has sometimes caused large trees to topple. Many orchids are extremely small and in some rich habitats there can be many different species on a single tree.

Most orchids are extremely long-lived in nature, and under favourable conditions in cultivation. In the Lankester Garden in Costa Rica, many native orchids flourish, but some of the largest plants are of Indian species that were imported into Costa Rica by Charles Lankester in the 1920's. Of the 4,000 or so, different species in the collection at the Royal Botanic Gardens, Kew, more than 500 have been there for more than 30 years. One plant from Central America has remained in the collection for more than 100 years, despite several moves from one greenhouse to another and other vicissitudes. But the life of many wild-collected orchids in cultivation is less than ten years. Sometimes it is only a few months.

Participants at this Congress would agree that it is the living collection of plants that is the heart of a botanic garden. While the scientists of tomorrow have to be educated and perhaps attracted to the professions of botany and conservation, it is the living plants in a garden that the visitors, young and old, come to see. Large and attractive orchid flowers are an essential feature of displays of tropical plants in many botanic gardens around the world.

The living plants are also essential material for research, for taxonomy, cytology, anatomy, biochemistry, and more recently, molecular biology and studies of breeding systems. An example at the Royal Botanic Gardens, Kew, is the collection of *Cymbidium* species which has been greatly enlarged in the last ten years. My colleagues David Du Puy and Phillip Cribb have found the living plants invaluable to elucidate many puzzling features for their recently published monograph of the genus. Anatomy of the leaves, chromosome studies, and fine details of flower and seed structure were easily studied in the living plants. At the same time as providing material for research, these plants are also attractive for the educational displays at Kew.

Orchids in botanic gardens should also provide material for propagation and *ex situ* conservation. My colleague Robert Mitchell describes the exciting work at Kew with the European orchids in a separate paper (Mitchell, this volume) Growing orchids from seeds is a micropropagation technique, because the seeds are so small and simple. It is a complicated procedure but it is not difficult. Anyone can do it, with care, even in the kitchen, with just a pressure cooker, some bleach and a plastic bag. With more sophisticated equipment it is much easier, and the results are more certain, but equipping a laboratory in order to raise orchids from seeds need not be expensive. Some botanic gardens have been raising flasks of orchid seedlings for many years. Germination experiments started in Singapore by the late Professor Eric Holttum in the 1930's have continued ever since and laid the foundations for the orchid cut-flower industry which is now so important economically throughout southeast Asia.

Vandas and dendrobiums bred in the Singapore Botanic Garden and in Hawaii can be grown easily in tropical climates with the minimum of attention and flower prolifically. Their propagation, harvesting as plants or as cut flowers, and sale, can be a useful source of funds if carefully managed. The development and display of colourful hybrids means that wild orchid species are no longer of great interest to the majority of amateur and commercial growers. Awards from Horticultural Societies, and a high income from sales, can be realised best from the newest man-made hybrids. Thus the orchid industry is to a very large extent of great, if indirect, assistance to conservation.

There is one orchid genus which is of special concern. The genus *Paphiopedilum* is very popular, because of its ease of cultivation and readiness to flower in many glasshouses. Some species have been grown in cultivation, both by amateurs and in botanic gardens, for more than 100 years. A few of the 60 or so species are now very rare in the wild while others are still collected in very large numbers, both for local and international trade. In China a number of new species have been discovered in the last ten years, and plants have been exported in great numbers, perhaps more than the populations can stand. Yet paphiopedilums can be raised from seeds very easily. It takes just a few years for a seedling plant to reach flowering size. As with the vandas and dendrobiums, there are more hybrid paphiopedilums than species in cultivation today. A *Paphiopedilum* hybrid was, in fact, one of the earliest of the man-made orchid hybrids in 1859. At the Royal Botanic Gardens, Kew, there is an expanding collection of hybrids of these slipper orchids. They are used for display to the public and as cut flowers for formal decorations, because the colours are brighter and the flowers last longer than those of many of the species. There is a wave of feeling among conservationists that the wild *Paphiopedilum* species must be protected by transferring the genus from Appendix II of CITES, where all except ten orchids are currently listed to Appendix I. This would effectively prevent all international movement of plants of this genus, both species and hybrids. Thus orchid specialists fear their efforts to encourage propagation from seeds, of the species as well as the hybrids, will put a risk if these restrictions on the international movement of plants materialise. Botanic gardens, whose staff often supply advice to CITES management authorities, have a responsiblity to think carefully about the possible consequences of such a move.[1]

This brief presentation has demonstrated some of the diversity and fascination of orchids. It is a family which surely must be included in any representative collection of flowering plants in botanic gardens. But for any one garden to obtain and maintain a comprehensive collection is clearly impossible and probably undesirable.

Many orchids are rare plants, very restricted in their distribution. Their ecological requirements, especially with respect to altitude, and thus temperature, are often very narrow. Thus the cultural conditions they need are demanding. Botanic gardens must therefore decide on their aims with respect to orchids and in relation to that decision they must bear in mind what facilities they can afford to provide and maintain. Specially trained or experienced staff are not easily found and kept. Suitable buildings with environmental controls are not cheap to build or maintain, either in the tropics or in temperate gardens.

---

1. At the Conference of the Parties in Lausanne, October 1989, the genus *Paphiopedilum* was transferred from Appendix II to Appendix I, as anticipated. This has already had the unfortunate effect, among overzealous officials, of inhibiting trade in flasked seedlings and other laboratory-raised plants of this genus, both species and hybrids, between many countries. It is hoped that the Secretariat in Lausanne will soon be able to encourage all management authorities to adhere strictly to the articles of the Convention. In the present circumstances it would appear that the Parties to CITES and conservationists in botanic gardens of many of the member countries are working at cross purposes.

At Kew there are nine growing areas for orchids which are heated, shaded, ventilated and humidified, as needed, throughout the year. In Singapore many orchids grow in the garden, in sun or shade as needed, but there is also an expensive greenhouse which is cooled and air-conditioned continuously for high altitude species.

The cost of maintaining an orchid collection need not be high if care is taken in its selection and presentation. At the same time, however, botanic gardens should set a good example to the public in their acquisition of plants. Most should be raised from seeds or obtained by exchange from other collections. The international movement of pollen, seeds and seedlings in flasks, is specifically exempt from the CITES permit system, so it is easy to arrange. The distribution of surplus seedlings and plants to other botanic gardens and to the enthusiastic public and even commercial growers, is a conservation measure that should be practised more often.

More involvement with amateur and commercial growers in education, display and the improvement of cultural techniques is highly desirable. Some botanic gardens involve local enthusiasts as voluntary helpers. The Marie Selby Botanic Garden at Sarasota, Florida, uses voluntary help for much of the repotting and record-keeping of its large collection of orchids.

The involvement and encouragement of amateur orchid growers can also provide a useful and sometimes unexpected source of funds. All the expenses of my Fellowship at Kew and the post occupied by Robert Mitchell, who is known as the Sainsbury Orchid Conservation Officer, have been provided by donations from two orchid enthusiasts. Sir Robert and Lady Sainsbury have supported our work, salaries and travels for several years. They have also paid for a large new growing house for orchids at Kew, and for computerised environmental controls in all the orchid houses. This beneficient relationship with Kew began when they presented their small private collection of orchids to Kew, in 1980, and their subsequent visits to check on the welfare of the plants were welcomed and encouraged. We are extremely grateful to sir Robert and Lisa Sainsbury for their continuing support.[2]

2. In August 1989, Sir Robert and Lady Sainsbury gave Kew a further £1 million as an endowment to make permanent the two posts they have supported up to now and to assist with the financing of other orchid work.

Botanic gardens have an important opportunity to play the leading rôle in many aspects of orchids research and conservation. I would encourage them all to regard an orchid collection, however small, as a necessary feature of their contents. They must become aware of and learn to exploit fully the tremendous interest that exists in the orchid family today. If the panda is the symbol of saving wildlife, orchids must surely be 'flagship' plants for habitat conservation.

The rôle of botanical gardens in orchid research

H.N. Rasmussen and F.N. Rasmussen

University of Copenhagen
Botanical Laboratory
Copenhagen, Denmark

Summary

Cultivated plants give the botanist an opportunity to study different life phases with descriptive as well as experimental methods and with a wide range of techniques, and to compare many taxa within a limited space. The complexity of orchid biology makes it particularly necessary to have orchid collections for supporting the research. On the other hand, this research will be helpful in setting a policy for the collection, and studies on the physiology of growth and reproduction are useful in the practical maintenance of the collection.

Plant material for botanical research can be divided in three categories: (1) Naturally occurring living plants that can be studied in the field, (2) living specimens, cultivated in a garden, and (3) dead and preserved herbarium specimens. There are advantages and drawbacks to all these kinds of study material. The herbarium specimen is usually a plant in flower, less often in fruiting stage. Vegetative herbarium specimens are often unidentified, and seedlings hardly ever occur in herbaria. Preserved specimens therefore tend to give a static image of the species in question. The cultivated plant can be followed through all seasons, and, provided it is propagated in cultivation, all stages in life, thereby giving a dynamic information of the lifecycle. The same applies to the field specimen provided that the access to the site is easy and regular. Very often, however, the botanist visits the site only once in a brief observation period. Unless a comparison between individuals of

TROPICAL BOTANIC GARDENS: THEIR ROLE
IN CONVERVATION AND DEVELOPMENT
ISBN 0—12—346850—7

different age and stages on the spot allows inference of the life cycle, the biological information gained from a momentary observation is essentially a static impression. In the case of herbarium specimens, information on the environment in which the plant occurred is usually restricted to brief notes made by the collector. In field material the environment is intact, including all interactions with other living organisms (predators, pollinators and so on) and can be studied and described in any level of detail desired. In contrast, the cultivated plant is virtually isolated from natural interactions with other species and placed in an environment which is artificial. This is a disadvantage, but at the same time entails unique possibilities: various environmental factors can be manipulated one by one and the effects on the plants observed. Major parts of classical plant physiology are based on such controlled experiments with whole plants in culture. Experiments with field material are also possible and indeed performed, particularly in pollination studies, but they are subjected to a variety of unpredictable conditions that complicate the interpretation of the results. The herbarium specimen, whether dried and pressed or pickled, is a durable material with rather small maintenance costs. Observations on preserved material can be repeated at a later occasion or by a different observer. The possibility of repeating the study exactly is a major scientific advantage. Data can be checked for errors, human or due to instrumentation. Studies based on garden specimens can in principle also be repeated, but living plants change with time and many plants naturally have a limited life span. The maintenance of a living collection is a skilled and time-consuming task and in spite of all efforts, certain losses regularly occur in a living collection. On the other hand, the field material is subject to many kinds of man-made and natural changes in the vegetation. Only in extreme cases can the study involve the same individuals repeatedly. The ease of access to the material determines the kind of equipment and the methods of analysis that can be applied to the study. In this respect the herbarium and cultivated material are ideal in being close to a laboratory and scientific staff, whereas the field material usually is studied with transportable and, therefore, less sophisticated instruments and methods.

To summarize the advantages of the garden collection, the cultivated plants offer the opportunity (a) to study the different life phases and seasonal changes provided these remain intact in culture, (b) to approach these with descriptive as well as experimental methods, and (c) to make comparisons between taxa within a limited space, which (d) is easily accessible for botanist and equipment, thereby enabling a wide range of techniques to be employed.

For ten years a series of orchid studies have been performed by staff members and students at the Botanical Laboratory of the University of Copenhagen, mainly on the basis of living material from the Botanical Garden in Copenhagen. These studies may serve as an example of the possibilities offered by an excellent living collection.

Taxonomic research

The core of the living orchid collection is the South East Asian orchids collected by Dr Gunnar Seidenfaden and originally accumulated for floristic and taxonomic work. When at its largest, this collection comprised about 3,000 living specimens and was very well kept and labeled. The regular picklings from flowering specimens were included in the spirit collection that today comprises about 10,000 specimens. Gunnar Seidenfaden's monumental work on the orchid flora of Thailand is well-known. The series of Orchid Genera in Thailand is completed (Seidenfaden, 1988) and a complete revision of Holttum's Orchids of Malaya (1964) is in progress (Seidenfaden & Wood, in prep.). Since the themes of orchid taxonomy and floristics are treated elsewhere in this volume they will be passed over very briefly here. However, the colour slide collection produced by the ecologist and eminent photographer Gsta Kjellson for Dr Gunnar Seidenfaden should be mentioned. The exposures were made during regular visits to the greenhouses over several years. Only with great difficulties could similar pictures have been taken in the field, since they represent rare and often widely dispersed species (cf., for instance, the colour plates in Seidenfaden, 1988). Flower colours are useful characters in many plant groups, but the diversity of orchid species and the intricate pollination syndromes pertaining to them make flower colours particularly needed in the description of orchid species. These colours are completely lost both in pickled and pressed flowers. In addition to the Thai orchid collection we have minor collections of temperate and subtropical terrestrial orchids, and tropical species from South American and Africa. These collections are essential for comparative studies covering the whole orchid family.

Phylogenetic research

Ontogenetic studies employ a number of laboratory techniques (sectioning, microscopy, photomicrography etc.) and are hardly possible without a cultivated collection. Material can be collected and fixed in the field, but this often requires a long stay in one place. It is easier to follow the development of a cultivated specimen and obtain study material for microscopic study at specified times. Phylogenetic research furthermore often requires a comparison between taxa from several geographic areas.

A number of studies have investigated the vegetative parts of the cultivated orchids. One example is the study of silica cells that are found in many orchids along supporting tissue. There are two distinct kinds. Some are spherical, shaped like a spiked mace, others are conical (Fig. 5-6 in Dahl Moller & Rasmussen, 1984). It is hard to understand why many epiphytic orchids accumulate silica and why many terrestrials do not, although the latter would seem much more exposed to silica in their environment (H. Rasmussen, 1986). Different species growing beside each other in the greenhouse may differ in respect to the kind of silica bodies produced. Since they are watered and fed in a standardized manner, we have in fact here done an experiment which demonstrates that silica cells are a genetically determined feature that is not modified by the environment. The available material showed an interesting distribution among major orchid groups, which supported the phylogenetic hypothesis (Fig. 1) that the conical silica bodies were present in the ancestors to recent Orchidales. Spherical bodies originate within the group, and has probably evolved twice (Dahl Moller & Rasmussen, 1984). Many orchids conform closely to a modular pattern of growth which is retained in cultivation. This fact can be ascertained by comparing cultivated plants with herbarium specimens of field collections. Again, the potential of the greenhouse material for interspecific comparisons is essential. The analysis of atypical branching patterns and densely congested shoot systems requires the preparation of fresh material under a dissection microscope, as well as the microscopic study of serial microtome sections. In this way a range of habits were found in cultivated specimens all referred to the species *Eria pubescens* (Andersen et al., 1986). This diversity was interpreted as variations on a common theme, that regulate the time of appearance of leaves and flowers in different ways, possibly related to different seasonal rhythms (Fig. 2). The organography of orchid flowers is more complicated than other flowers. Styles and stamens are fused into a compound structure, the column, often with intricate details that are essential in the pollination of the flowers (F. N. Rasmussen, 1986a). The orchid column is a sufficiently complicated organ to allow phylogenetic inference to be drawn

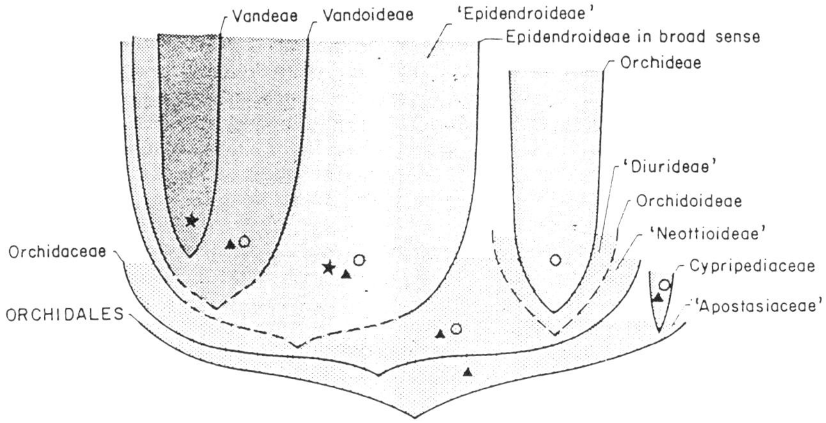

Figure 1. A tentative scheme of clades in Orchidales (from Rasmussen, 1985) is used to show the occurrence of the two kinds of silica bodies. Broken lines indicate less corroborated clades. Names of clades are indicated at the periphery of the groups, while names of paraphyletic groups are indicated pointing toward the interior of the clades to which they belong and given in quotation marks in accordance with Wiley's conventions of phylogenetic classification (Wiley, 1981). Several lower-order clades, are omitted for the sake of clarity.

: conical silica bodies,

: spherical,

o : lack of silica bodies.

Conical bodies occur in most orchid groups and were probably present in ancestral orchids; the spherical bodies arose within Epidendroideae, and may have evolved once more, in the ancestor to Vandeae. (from Dahl Moller & Rasmussen, 1984).

from additions or deletions along the ontogenetic pathway (Fig. 3, F.N. Rasmussen, 1986b).

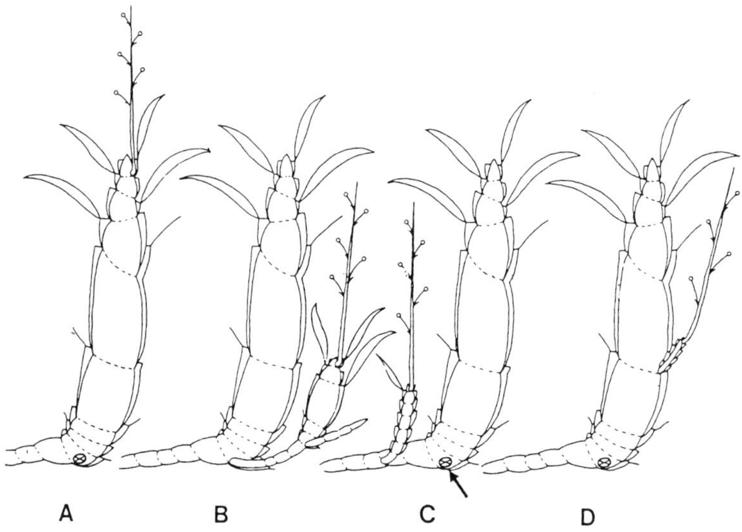

Figure 2. Range of habits, represented by different cultivated specimens belonging to *Eria pubescens*. A. Inflorescences develop laterally on pseudobulbous shoots, that are all alike. B. Alternating vegetative modules with large foliage leaves, and floriferous modules, that are short, with small or no foliage leaves, and with lateral inflorescences. C. Floriferous modules are issued from a node below the renewal bud (arrow) and therefore does not take part in the continuation of the sympodial rhizome. D. Floriferous module produced above the renewal bud, recognizable as a module by the scale leaves at base, and thereby being distinguishable from A. (from Andersen et al., 1988).

Pollination biology

We do not know enough about the precise function of the diversity of columns found in orchid flowers, but an exact knowledge of the column structure is fundamental for understanding the the precision with which pollen are placed on the pollinators. Mechanisms that can be imagined on the basis of a detailed morphological study can lead to the formulation of hypotheses that can be tested in field studies of pollination (such as Kjellson, Rasmussen, & Dupuy, 1985). Pollination syndromes in orchid species are usually studied in the field where the behavior of pollinators can be observed. On the other hand, artificial pollinations among cultivated plants can also provide important information on the reproduction biology, perhaps of a more fundamental nature than case studies in the field could give.

The compatibility system of the large genus *Dendrobium* was investigated recently on the basis of about 1,700 pollination experiments, involving 62 species grown in the greenhouses (Johansen, 1990). Most species proved to be self sterile and to reject their own pollen by

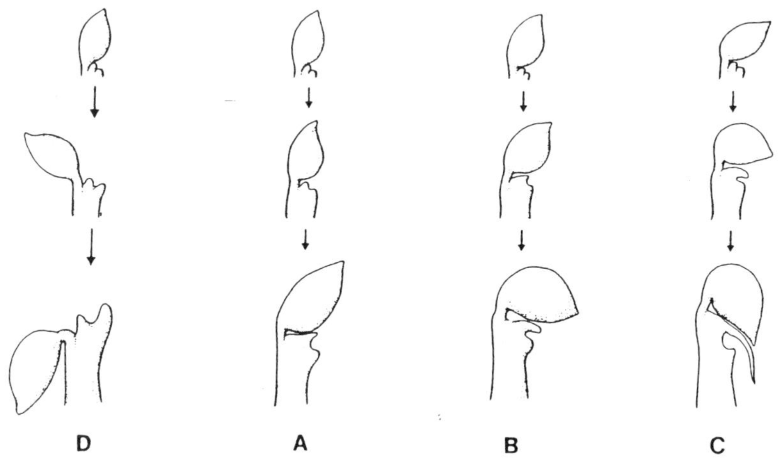

Figure 3. Typical development of the column of the flower in four subfamilies of Orchidaceae. A: Neottioideae, B: Epidendroideae, C: Vandoideae, D: Orchidoideae. The final stages in A and B are reached as intermediate stages in the more advanced ontogenies of B and C, respectively. The ontogeny of D takes a different course resulting in a recumbent anther. The observations were made on dissected and sectioned buds from plants cultivated in various botanical gardens (Rasmussen, 1986b).

mechanisms that are unique. Orchid pollen is aggregated in so-called pollinia which are transferred to the stigma at pollination. Each pollinium contains a number of pollen grains which corresponds to the huge number of seeds produced in the fruit. After a pollination with compatible pollen, the pollen tubes slowly grow down the stylar canal and into the ovary, eventually to reach the ovules and fertilize them. Depending on the species, this process takes from a few weeks to many months and during this time the ovules differentiate from the placental tissue (Fig. 4). After an incompatible pollination, however, the flower including the inferior ovary is shed right after the application of the pollen to the stigma. These peculiar facts have been known since the time of Hildebrand (1863). The orchid stigma is filled with a pectinaceous slime in which free-living cells are floating (Hildebrand, 1863, Calder & Slater, 1985, Johansen, 1990). Such a suspension of living higher plant cells is very rare as a natural phenomenon, although it is often produced *in vitro* with tissue culture techniques. Applications of dead pollen, mentor pollen and other experiments support the idea that a reaction takes place between the pollen and these loose stigma cells (Fig. 5), so-called eleutherocytes (Johansen, 1990). Eleutherocytes have an ultrastructure indicating a high secreting activity (Slater & Calder, 1990, Johansen, MS). As a strong indication of their involvement in the compatibility reactions there is now convincing evidence that the structure of these cells differs according to the type of pollen applied (Johansen, in prep). Eleutherocytes can be cultivated on an artificial medium and are presently being subjected to various *in vitro* experiments at the Botanical Laboratory (Johansen, in prep.).

Vegetative and sexual reproduction

Apart from the usefulness of orchid collections to fundamental orchid research, a collection usually also benefits from the research that takes place around it. All information on the biology of the cultivated species is helpful for practical cultivation as well as for formulating the policy for the collection. The research will help to define the cases where it is essential to keep many specimens of a species, and to select the species and genera to be acquired, collected, kept, and propagated. It will be an increasing obligation for living botanical collections to propagate material once it is acquired, instead of renewing the collection with specimens obtained in nature, particularly in the case of threatened species.

Many orchid species are rather easy to propagate, either by seeds or cloning. Large-scale propagation of identical offspring is usually not the issue for scientific collections, but vegetative propagation *in vitro* may be feasible in species with an architecture unsuitable for divisions, such as the monopodial orchids (Vandeae). In a series of experiments with vegetative micropropagation we have exploited the fact that the ovary does not contain any gametophytic (haploid) tissue before pollination (Rasmussen & Rasmussen, in prep.). The carpels of unpollinated flowers can be induced to develop protocorm like bodies (resembling orchid seedlings) *in vitro* (Fig. 6-7). Young flower buds are removed with no harmful effects to the donor plant, surface-sterilized and planted with the ovary in sterile growth media rich in sucrose and auxin. Organogenesis beyond the protocorm stage takes place after transfer to a hormone-free substrate. We are presently investigating the range of species to which the method can be applied. However, the extent to which the hormone treatments could induce genetical abnormalities is not known.

With respect to preserving the genetical range in the cultivated species, propagation by seeds evidently is much preferable to cloning. Considering the minute size of orchid seeds, seed banking could have enormous potential for species conservation but reliable methods for viability testing and germination are necessary. A preservation of a high germination percentage is desirable to ensure that the storage regime does not subject the seed population to a strong selection pressure. The well-known association of orchid seedlings with mycorrhizal fungi complicates the investigation of seed physiology because the lack of a suitable fungus or an appropriate chemical substitution for it can be invoked whenever orchid seeds fail to germinate *in vitro*. Viability testing of the embryos by staining is possible, although difficult because it requires the removal of the testa (Pritchard, 1985).

Some orchid species, particularly the terrestrials, have very complex requirements for germination and some of these requirement may not have been identified yet. No single chemical factor seems responsible for the succesful asymbiotic germination of orchids (Arditti, 1967). The germination response of easily germinated terrestrial species varies markedly with temperature, light, acidity of the medium and other

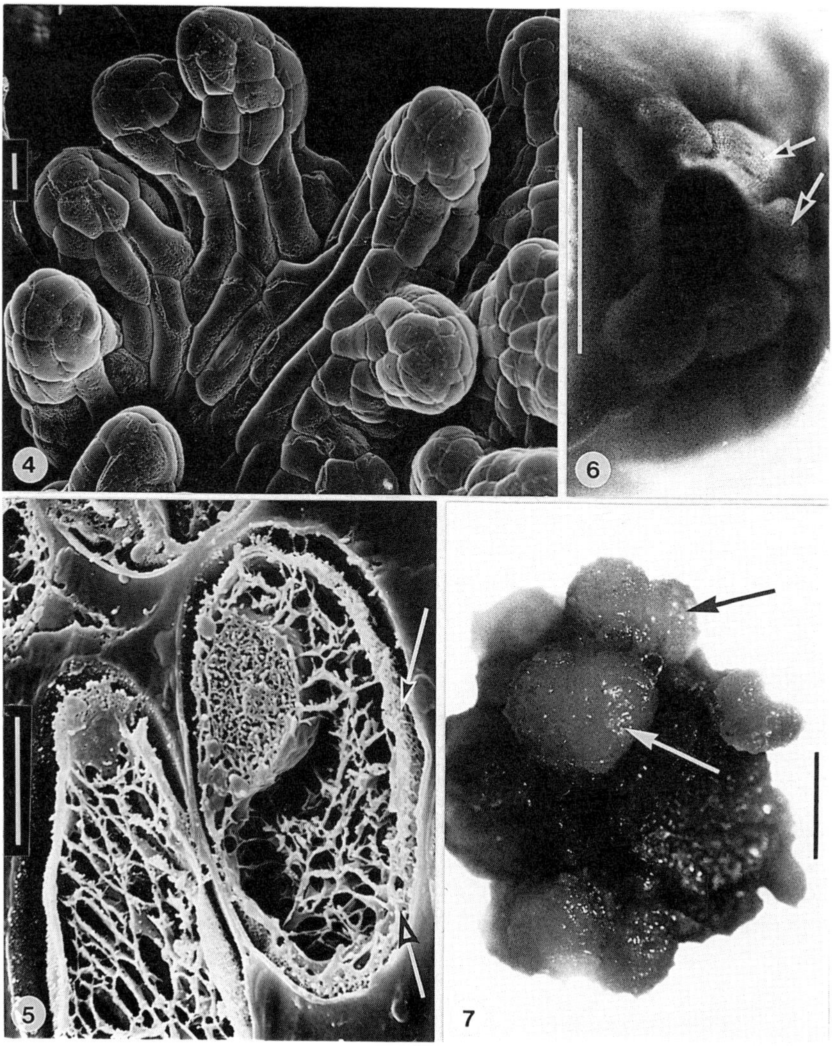

Figure 4. The placenta of an orchid fruit (*Dendrobium secundum*) 17 days after pollination. While the pollen tubes are already present in the ovary, the ovules develop from these lobes, and will subsequently become surrounded by integuments, produce a female gametophyte and become fertilized. The maturation of the seeds is completed 51-97 days after pollination. Freeze scanning electron microscopy of fresh material. Scale bar: 10 m. By courtesy of B. Johansen.

Figure 5. *Eleutherocytes* are freely suspended, living cells that float in the stigmatic cavity of orchid flowers. The cells are fractured to show the cellular contents and the finely undulated plasmalemma (arrows), indicative of a secretory activity. Freeze scanning electron microscopy, fresh material. Scale bar: 10 m. By courtesy of B. Johansen.
Figure 6. The insertion of the inferior ovary of a flower bud into a growth medium may induce the carpels to callusing. The proliferation in particular comes from the outer surface of the carpels (arrows). *Pholidota imbricata*, 22 days of culture, MS- medium with 5.0 mg NAA l-1, 20 g sucrose l-1. Scale bar: 1 mm. Orig.

Figure 7. The callus organizing into protocorm-like bodies (arrows) which upon transfer to hormone-free substrate will start organogenesis. *Pholidota inbricata*, one year old culture, MS- medium with 2.0 mg NAA, 2.4 mg BA, and 20 g sucrose l-1. Scale bar: 1 mm. Orig.

factors, in spite of the presence of an undoubtedly compatible fungus (Rasmussen et al., 1990a, 1990b, and unpublished data). Therefore it is possible that unfavourable conditions induce or maintain a dormancy in the seeds which is not broken by a compatible fungus. Decrease in germinability after storage of the seeds could be due to induced dormancy as well as loss of viability. We are presently investigating the responses of seeds to such factors as molarity and redox potential of the substrate, sucrose concentration, fluctuating temperatures, and composition of the atmosphere. By describing the requirements of different fractions of the seed population in easily germinated species we hope to get clues to the germination of the difficult species.

Conclusions

It should be clear from the examples mentioned above that living orchid collections have much to contribute to the fundamental research in orchid biology. Furthermore, the practical requirements for maintaining a living collection may inspire applied research in germplasm storage and reproduction physiology. Other laboratories associated with well established living collections could provide similar examples.

Orchids hold a special place in the plant kingdom, providing fascination for scientists as well as the general public. Orchids are biologically outstanding to an extent we are only beginning to realize. The symbiosis of orchid seedlings with fungi, which is more like a highly adapted parasitism, is receiving new scientific attention and many new facts will emerge about this unique relationship in connection with its practical application in plant production (Mitchell, this volume). There are many puzzles to solve in specialized structures of the vegetative parts, particularly of the epiphytes, and intriguing shapes and patterns in the flowers which presumably often serve to exploit and cheat pollinators. The extent of parallel evolution exhibited in these structures is so conspicuous that they seem to defy the traditional neo-Darwinian set of evolutionary hypotheses. Furthermore, new data indicate that orchid have a unique mechanism to avoiding self-pollination by cell-to-cell interactions between pollen and a cell suspension in the stylar canal. One might say that the more we learn about orchid biology the more rewarding and necessary the research in the living collections seem to become.

References

Andersen, T.F., Johansen, B., Lund, I, Rasmussen, F.N., Rasmussen, H. & Srensen, I. (1988). Vegetative architecture of Eria. *Lindleyana* 3, 117-132.

Arditti, J. (1967). Factors affecting the germination of orchid seeds. *Botanical Review* 33, 1-97.

Calder, D.M. & Slater, A.T. (1985). The stigma of *Dendrobium speciosum* Sm. (Orchidaceae): A new stigma type comprising detached cells within a mucilaginous matrix. *Annals of Botany* 55, 297-307.

Dahl Moller, J. & Rasmussen, H.. (1984). Stegmata in Orchidales: character state distribution and polarity. *Botanical Journal of the Linnean Society* 89, 53-76.

Hildebrand, F. (1863). Die Fruchbildung der Orchideen, ein Beweiss fr dis doppelte Wirkung des Pollens. *Botanische Zeitung* 21, 329- 333, 337-345.

Holttum, R.E. (1964). *A revised flora of Malaya*. Vol 1. *Orchids of Malaya* (ed.3) Government Printing Office, Singapore.

Johansen, B. (1990). Incompatibility in *Dendrobium* (Orchidaceae). *Botanical Journal of the Linnean Society* 103, in press.

Johansen, B. (199?). Ultrastructure of the unpollinated receptive stigma in *Dendrobium lindleyi* Steud. (Orchidaceae). Planta.

Kjellsson, G., Rasmussen, F.N., & Dupuy, D. (1985). Pollination of *Dendrobium infundibulum*, *Cymbidium insigne* (Orchidaceae) and *Rhododendron lyi* (Ericaceae) by *Bombus eximius* (Apidae) in Thailand: a possible case of floral mimicry. *Journal of Tropical Ecology* 1, 289-302.

Pritchard, H.W. (1985). Determination of orchid seed viability using fluorescein diacetate. *Plant, Cell and Environment* 8, 727- 730.

Rasmussen, F. & N.(1985). Orchids. In *The families of the monocotyledons: structure, evolution, and taxonomy*. R.M.T. Dahlgren, H. T. Clifford, & P.F. Yeo (eds), pp. 249-274. Springer, Berlin.

Rasmussen, F.N. (1986a). On the various contrivances by which pollinia are attached to viscidia. *Lindleyana* 1, 21-32.

Rasmussen, F.N. (1986b). Ontogeny and phylogeny in Orchidaceae. *Lindleyana* 1, 114-124.

Rasmussen, H. (1986). An aspect of orchid anatomy and adaptationism. *Lindleyana* 1, 102-107.

Rasmussen, H., Andersen, T.F. & Johansen, B. (1990a). Temperature sensitivity of in vitro germination and seedling development of *Dactylorhiza majalis* (Orchidaceae) with and without a mycorrhizal fungus. *Plant, Cell and Environment*, 13, 171-177.

Rasmussen, H., Andersen, T.F. & Johansen, B. (1990b). Light stimulation and darkness requirement for the symbiotic germination of *Dactylorhiza majalis* (Orchidaceae) in vitro. *Physiologia Plantarum*, 78, in press.

Seidenfaden, G. (1988). Orchid Genera in Thailand 14. Fifty-nine vandoid genera. *Opera Botanica* 95, 1-398.

Slater A.T. & Calder, D.M. (1990). Fine structure of the wet, detached cell stigma of the orchid *Dendrobium speciosum* Sm. *Sexual Plant Reproduction* 3, 61-69.

Wiley, E.O. (1981). *Phylogenetics*. Wiley & Sons, New York.

Orchid propagation in botanic gardens

Robert B. Mitchell*

Sainsbury Orchid Project
Royal Botanic Gardens
Kew, Surrey, U.K.

Summary

Epiphytic and terrestrial orchids are now successfully raised in large numbers from seed for research, horticulture and conservation. Even the European terrestrial orchids, for long thought to be difficult to raise, have been successfully propagated from seeds and re-introduced to the wild by the Sainsbury Orchid Conservation Project at the Royal Botanic Gardens, Kew.

Introduction

The earliest attempts of orchid propagation from seeds involved sowing fresh seeds around the roots of parent plants. This method is still used today by orchid specialists and enthusiasts, (e.g. Vogelpoel 1983 for *Disa* spp.), although the results are often unreliable. The discovery that orchid seeds germinate naturally with the stimulus of the mycorrhizal fungus was made by two scientists, Bernard (1902, 1909) and Burgeff (1909). This led to the development of a technique for germinating orchid seeds using the fungi isolated from the roots of parent plants. As early as 1906, the 'pure culture method' or 'symbiotic method' was being successfully used on a commercial scale by Joseph Charlesworth, in England, for the germination and growth of *Odontoglossum* and allied genera (Ramsbottom 1923). Knudson (1922) developed a nutrient medium that could substitute the need for the mycorrhizal fungus. This

*Present address: c/o The Cottage, Church Road, Coddenham, Ipswich IP6 9PY, U.K.

TROPICAL BOTANIC GARDENS: THEIR ROLE
IN CONVERVATION AND DEVELOPMENT
ISBN 0–12–346850–7

medium contained sucrose and basic plant nutrients in the form of mineral salts. This method proved to be particularly useful for propagating tropical epiphytic orchids and their hybrids. Now, with further advances in nutrient media, the great majority of orchids can be raised using this 'asymbiotic' technique.

Orchid propagation at the Royal Botanic Gardens, Kew.

Orchid propagation from seeds has become an important method of helping to maintain and enhance the living orchid collection at the Royal Botanic Gardens, Kew. The collections are extensively used for scientific studies and conservation purposes as well as display and education. In the 1960s, Dr Peter Thompson, then studying seed physiology at Kew, began to take an interest in germination requirements of orchid seeds. Dr Thompson's results encouraged the Gardens to expand the work as part of a micro-propagation unit. The unit uses *in vitro* techniques to raise plants for the collections that had proved to be difficult to multiply by conventional methods. The majority of orchids raised are tropical epiphytes and the asymbiotic technique is used exclusively. The staff at Kew have, to date, handled over 3,500 different orchid seed accessions since the establishment of the laboratory in 1974. Little work has been done on the vegetative propagation of individual species or clones *in vitro* as we prefer to maintain the greatest possible potential for genetic diversity within the collection. Vegetative propagation should not however be dismissed completely as it can help to increase desirable material for display and other uses.

Development of the collection

Raising orchids from seeds has the potential of becoming the primary source of living material for botanic garden collections. The vigorous seedlings which can be raised in large numbers allow the collection to hold a wider genetic sample of the species, within the limits of available space, and without the need to remove mature plants from their natural habitats. Establishment of wild source plants can be difficult even for very experienced growers, particularly in temperate climates where seasonal variations in daylength and temperature are a problem. Experienced growers have recorded a success rate of only 80% with wild-collected material. Greater use of seed-raised plants could help to reduce these losses as larger numbers of plants can be raised and weaned. The timing of removal of plants from culture and the weaning environment is directly controlled by the grower and thus higher success rates can be achieved. Plants can also be retained in culture so that if the seedlings do not establish, additional material is available.

Producing large numbers of plants

Propagation from seeds also has a role in increasing plant material needed for particular research projects, providing advance planning can give sufficient time to allow the plants to reach maturity if they are needed in flower. Botanic gardens need to be able to offer their visitors interesting and changing displays that excite the imagination. The natural beauty of orchids has a tremendous public appeal and bold displays of flowering plants are an invaluable addition to high intensity visitor areas.

Cross pollination is always the most desirable method for producing seeds. However, there are many instances where a single rare or valuable plant is held in the collection. Self pollination, by hand, usually allows the development of thousands of seeds from such plants so that large numbers of plants can be produced and seedlings can be distributed to other botanic gardens and reliable growers. Self pollination is not ideal but if carried out carefully it can ensure the survival of a species, at least in cultivation, without the need for recollection from the wild.

An additional benefit of production from seeds is the availability of surplus material for distribution and exchange. Distribution can take place at several different stages of production depending on where the material is being sent. Seeds are the cheapest material to distribute and can be sent to most countries of the world without quarantine problems. Sterile cultures of germinated seedlings at varying stages can be distributed to establishments with the facilities to grow them on *in vitro*. In most cases the best time for distribution is when the plants are at the optimum stage for weaning from culture. The plants are still sterile and thus do not contravene quarantine regulations of most countries, so there is little delay in their arrival. Seeds and seedlings in culture are specifically exempt from the import/export permit system of CITES which also facilitates their rapid transport internationally. The final stage for distribution is as weaned plants. Kew's experience of plant distribution to date has highlighted the importance of providing accurate data with the plant material. This may need to include cultural recommendations to help growers with less experience of the species.

All documentation for the plants e.g. CITES permits, plant quarantine permits and import permits should be sorted out before the request for plants is sent in. Feedback on the condition of the plants on arrival and success with their growth is very useful to the dispatchers as it helps to ensure that a plant distribution system achieves its goal.

Specific projects on endangered species

In 1983 a specific project was established at Kew to work on growing the European terrestrial orchids, many of which have proved to be difficult to germinate by asymbiotic methods and difficult to grow to flowering-sized plants. The Sainsbury Orchid Conservation Project was inspired by the work of Mark Clements (1982), at the Australian National Botanic Garden, Canberra. Clements had revived the use of the symbiotic technique to grow several species of Australian terrestrial orchids from seeds. Dr Phillip Cribb saw the value of this work in relation to British and European orchids and with the support of the Director and the generous backing of Sir Robert and Lady Sainsbury, invited Clements to Kew to initiate the project.

An ecological approach to growing these difficult plants has been applied throughout the projects history. This approach has proved to be very successful at each stage of the plants growth, from the initial sowing where a symbiotic fungus is introduced, through to the weaning and growing stages where the composts and cultural practices are designed to maintain the growth of the fungus in a balanced symbiosis with the plants. During the last two years the project has started planting trials in the open ground within the gardens and carefully selected wild localities. The first of these plantings, *Orchis laxiflora*, raised using the symbiotic method, surpassed our expectations. The majority of the plants not only survived the transition to the soil but a small proportion flowered as two year old seedlings. This trial has served as a role model for additional plantings of this species as well as new plantings of *Dactylorhiza fuchsii, Dactylorhiza praetermissa, Ophrys apifera, Orchis morio* and *Cypripedium calceolus*.

The most significant in conservation terms has been the progress with *Cypripedium calceolus*. This beautiful orchid has been over-collected to the extent that the sole surviving plant in Britain now requires a round-the-clock guard to ensure that it is not stolen or damaged. The initial hope that the symbiotic method would help to stimulate the germination of the mature seeds of *C. calceolus* has not yet given positive results. Many attempts have been made to find a mycorrhizal fungus, but so far none has been isolated.

Successful germination has been achieved using a technique first used by Withner (1943). In 1986 seeds from an immature capsule resulting from a hand pollination were sown onto a range of media. There was some germination and after six months there were nine seedlings growing at Kew. The seedlings proved to be much more demanding than most of the other terrestrial orchids we have grown and the initial seedlings declined. Contact with colleagues in Europe and America revealed that their experience with the plant was similar. The representatives of the genus *Cypripedium* held in the living collections were frequently disturbed during the course of the year to look for mycorrhizal fungi in their roots. Observation of these mature plants revealed the natural growing cycle. Once this pattern was more clearly understood, seedlings have been grown on to produce their first leaves and weaned from culture. The most recent progress is a trial planting of natural source material in a wild habitat, where the seedlings have come through the winter and the growing buds are emerging through the soil. To date the emphasis of the project's work has been largely conservation and horticultural, rather than scientific. This direction was taken because of the limited material available and the urgent need to ensure that this species is not lost from the British flora.

The work of the project on a specific conservation problem highlights the importance of the living collection as a research resource. A lot of information has been gained about the life cycles of the plants from their growth patterns in cultivation. This has enabled the project to give more accurate advice on the management of the remaining orchid populations in Britain. Information about the length of time needed for seed ripening allows more accurate management decisions to be made in natural habitats. Training workshops for wardens working with rare plants in natural habitats have helped increase the level of understanding of the plants for field studies and habitat management.

Communication

There is much to be gained from increasing the level of communication and exchange, both within and between botanic gardens, conservation bodies, orchid enthusiasts and commercial growers. The progress of the Sainsbury Orchid Conservation Project has been helped considerably by openly discussing techniques and problem areas with commercial and amateur growers, conservationists and other researchers at universities and botanic gardens. It should also be possible to arrange co-operative propagation strategies. Not all botanic gardens have the resources to set up their own propagation facility, with its requirement for specialist equipment and trained staff. Many gardens could co-operate with commercial growers or enthusiasts who welcome the opportunity to increase their stocks without removing plants from the wild. A good illustration of such a project is the cooperation between the Equatorial Plant Company, the Royal Botanic Garden, Edinburgh and the Scottish Orchid Society (H. Marsh et al. 1989). This Orchid Preservation scheme is attempting to use the skills and specialisms of the three organizations to maintain collections of the Oxyglossum section of *Dendrobium*. Positive encouragement from government and conservation bodies, in the form of easier trade in plants of known cultivated source, i.e. seed-raised or vegetatively propagated, would help considerably to reduce the 'Black Market' trade in rare and endangered species.

References

Bernard, N. 1903. La germination des Orchidées. *Compt. Rend. Acad. Sci. Paris* 137, 483-485

Bernard, N. 1909. L'évolution dans la symbiose les Orchidées et leur champignons commensaux. *Ann. Scie. Nat. Bot.* 9, 1-96.

Burgeff, H. 1909. *Die Wurzelpilze der Orchideen, ihre Kultur und ihr Leben in der Pflanze*. G. Fischer, Jena. 220 pp.

Clements, M.A. 1982. Development in the Symbiotic Germination of Australian Terrestrial Orchids. In: J. Stewart & N.C. van der Merwe, ed. *Proceedings of the 10th World Orchid Conference.* South African Orchid Council, Johannesburg.

Clements, M.A., Muir, H. and Cribb, P.J. 1986. A preliminary report on the symbiotic germination of European terrestrial orchids. *Kew Bulletin* 41(2), 437-445.

Knudson, L. 1922. Non-symbiotic germination of orchid seeds. *Bot. Gaz.* 73, 1-25.

Marsh, H., Warren, R. & Woods, P. 1989. Orchid Preservation. A practical proposition for orchid societies. *The Orchid Review* 97(115) (March 1989), 83-85.

Muir, H., 1987. A Symbiotic micropropagation of *Orchis laxiflora*. *The Orchid Review* (Jan. 1987), 27-29.

Ramsbottom, J. 1923. Orchid mycorrhiza. *Trans. Brit. Mycol. Soc.* 8, 28-61.

Vogelpoel, L. 1983. *Disa uniflora: Its propagation and cultivation.* Orchid Society of South Africa, Cape Town.

Withner, C., 1943. Ovule culture: A new method for starting orchid seedlings. *Am. Orch. Soc. Bull.* 11, 261-263.

Orchid conservation in Bogor Botanic Gardens and its associated gardens

S. Riswan, Irawati and Sukendar

Herbarium Bogoriense
Research and Development Center in Biology
Lipi, Bogor, Indonesia

Summary

Bogor Botanic Gardens (Kebun Raya Bogor) was established on May 18th by Dr C.G.J. Reinwardt. Formerly it was known as S'Land Plantentuin Buitenzorg or officially called the Hortus Botanicus Bogoriensis. This garden is the oldest botanic garden in Indonesia. There are three associated gardens established afterwards; Cibodas Garden (1874), Purwodadi Garden (1941) and Bali Garden (1959). In this paper, the history of the garden as an *ex situ* centre for conservation of orchids is discussed. It includes the fluctuation in size of its orchid collections, its rôles as source of knowledge in orchids and the problems of its development and maintenance in the future.

Introduction

Kebun Raya Bogor (Hortus Botanicus Bogoriensis) was established in 1817 by Prof. C.G.C Reinwardt, one year after the transfer of Java to the Dutch Government by the English in 1816. He was the first Director of this Botanic Garden. The Garden was founded with dual purposes of promoting botanical investigation in Indonesia and for experimentation on important crops. Bogor Botanic Garden is one of the oldest botanic gardens in the tropical region and the first and oldest botanic garden in Indonesia.

TROPICAL BOTANIC GARDENS: THEIR ROLE
IN CONVERVATION AND DEVELOPMENT
ISBN 0–12–346850–7

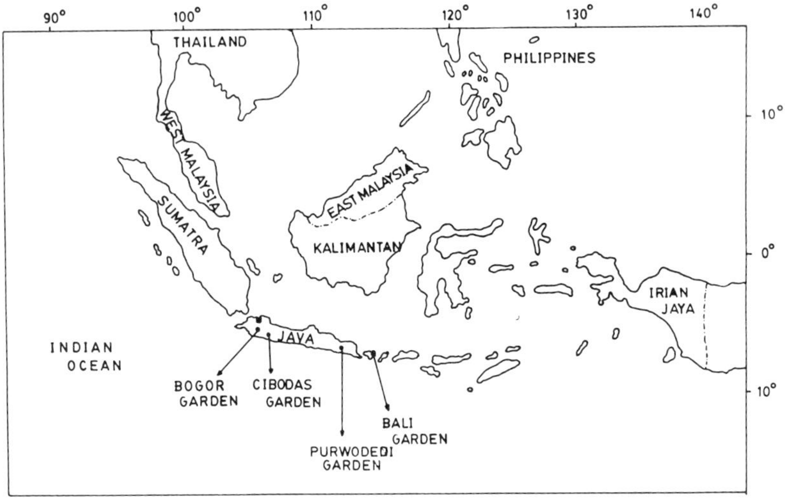

Fig . 1 : Map of Location Bogor Botanic Garden and its Associated Gardens

Table 1. Index Similarity* between Bogor Garden and Its Associated Gardens based on present and absent of Orchid collections

	BBG	CBG	PBG	BIBG
BBG	-			
CBG	18.39	-		
PBG	29.11	16.97	-	
BIBG	23.02	20.04	22.10	-

Legend: BBG = Bogor Botanic Garden
CBG = Cibodas Botanic Garden
PBG = Purwodadi Botanic Garden
BIBG = Bali Botanic Garden
* = Jaccard' Index Similarity

Three of the Bogor's associated gardens are Cibodas, Purwodadi and Bali, which were established in 1874, 1941 and 1959 respectively. The Cibodas and Bali Gardens are mountain gardens and the Purwodadi Garden in East Java is a lowland dry-monsoon. All these gardens (Fig 1) are under the administration of the Research and Development Center for Biology, Indonesian Institute of Sciences (LIPI).

In the last two decades in Indonesia, many tropical rain forests which are natural habitats of orchid plants have being disturbed. Most of them have been logged for timber. Others are converted into agricultural lands, mining and resettlement or transmigration areas, industrial estates and the networks of roads. Consequently, many wild orchid species will become extinct or severely eroded genetically unless preventive measure are taken. It means that *in situ* conservation is becoming increasingly more difficult. Wild population of orchids depends upon the stability of their natural habitats. When their habitats are disturbed they disappear. The problems are getting worse, since the collecting of wild and rare species for commercial purposes in their native habitats has increased.

The rôle of botanic gardens for *ex situ* conservation is therefore extremely very important a source of orchid germplasm. In this paper the history, problems and prospects of Bogor Botanic Gardens and its association gardens with regard to *ex situ* orchid conservation will be discussed.

The gardens and their orchid collections

Through time, there were fluctuations in the number of orchid species cultivated in the Bogor Garden and its associated gardens (Figures 2 to 5). The main reason for this was the inadequacy of the budget for maintenance and a shortage of qualified staff and technicians. Other reasons were vandalism and the theft of specimens. Many collections died because of carelessness. Other collections require re-identification and verification while others remain unnamed. In the last few years there has been little new exploration and collection for new accessions for the Bogor Garden and its associated gardens.

The Bogor Garden had its largest orchid collection in 1930 with 126 genera and 1,024 species. In 1982, the Cibodas Garden had 61 genera and 230 species), and in 1988, the Purwodadi Garden has 86 genera and

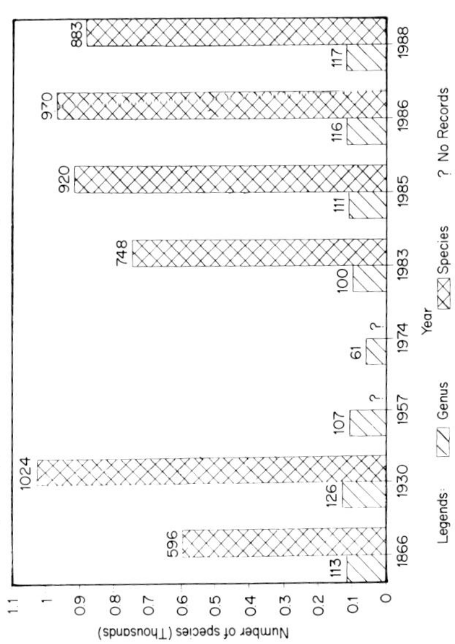

Fig 2 Histogram of orchid collection number at Bogor Botanic Gardens

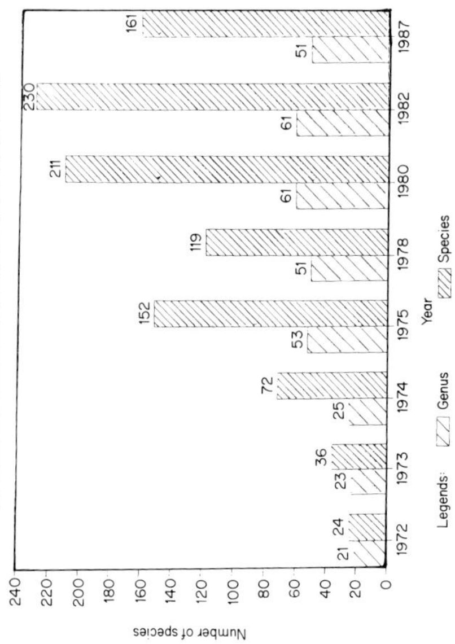

Fig 3 Histogram of orchid collection number at Cibodas Botanic Gardens

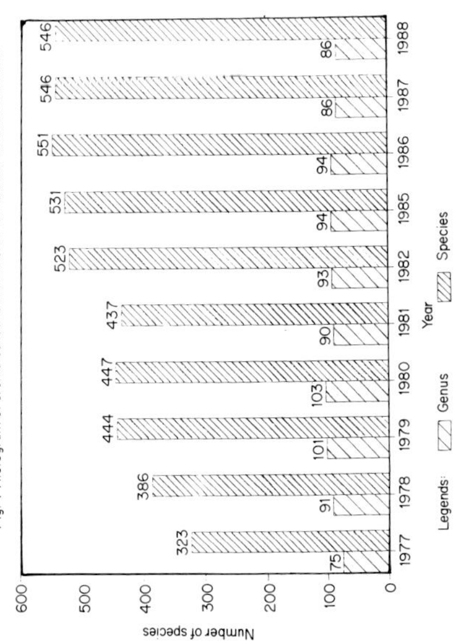

Fig 4 Histogram of orchid collection number at Purwodadi Botanic Gardens

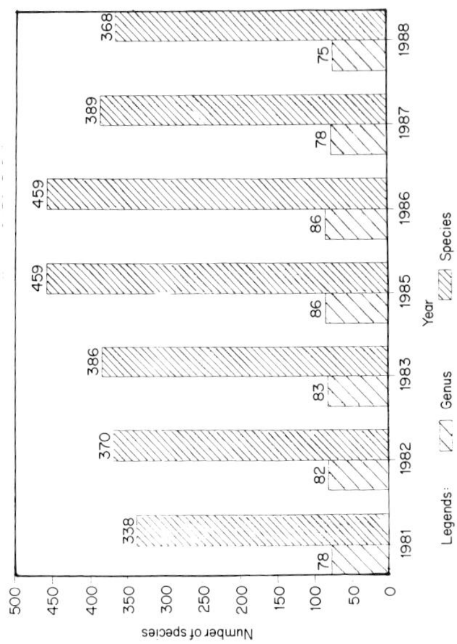

Fig 5 Histogram of orchid collection number at Bali Botanic Gardens

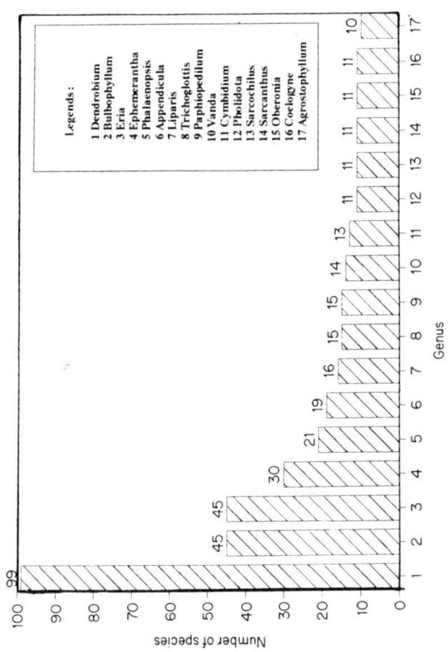

Fig. 6 Histogram of genus with number of species > 10 at Bogor Botanical Gardens

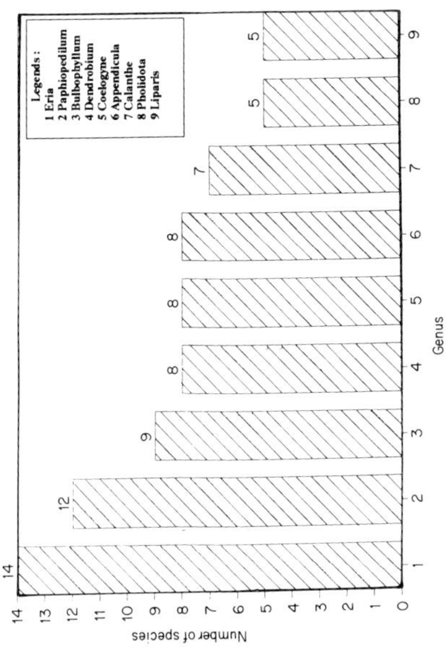

Fig. 7 Histogram of genus with number of species > 5 at Cibodas Botanical Gardens

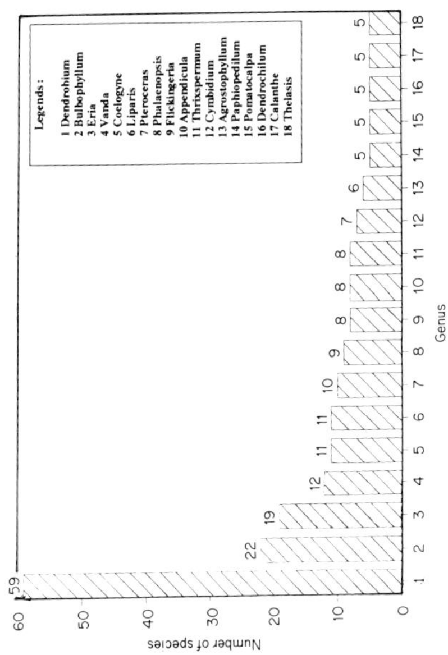

Fig. 8 Histogram of genus with number of species > 5 at Purwodadi Botanical Gardens

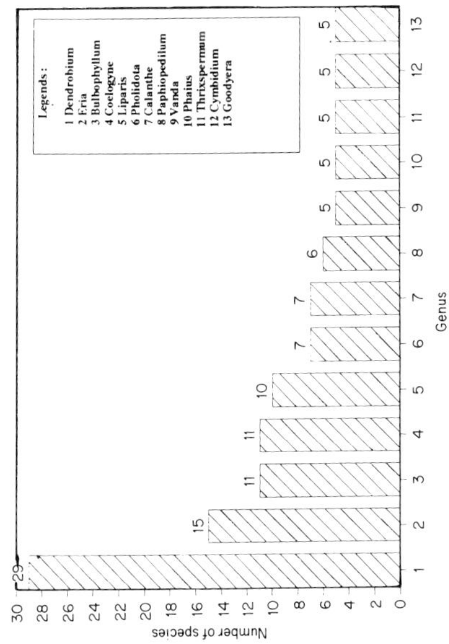

Fig. 9 Histogram of genus with number of species > 5 at Bali Botanical Gardens

546 species. Meanwhile, in Bali Garden, the largest collection was reached in the years 1985 and 1986 with 86 genera and 459 species.

In 1988, the Bogor Garden collection of orchids was the largest (117 genera and 883 species) and the smallest was that of the Cibodas Garden (51 genera and 161 species). Even though the Purwodadi and Bali Gardens are young compared to the Cibodas Garden their collections are greater as they received many additional collections gathered during their own field trips and as dublicates from the Bogor collections from the eastern part of Indonesia (West Irian, Mollucas and Lesser Suanda Islands).

Figures 6 to 9 show the most common orchid genera in Bogor Garden and its associated gardens. The most common genera (of which more than 10 species are cultivated) are 17 genera in the Bogor Garden, 2 genera in the Cibodas Garden, 7 genera in the Purwodadi Garden and 5 genera in the Bali Garden. They are *Dendrobium*, *Bulbophyllum*, *Eria*, *Coelogyne* and *Liparis* (Bogor, Purwodadi and Bali Gardens) and *Eria* and *Paphiopedilum* (Cibodas Garden). Bogor Garden has 17 species of the slipper orchid (*Paphiopedilum*) and the Cibodas Garden has 12 species. Even though, the other two gardens have only 5 or 6 species of this genus in cultivation they do grow the rarest species from Java, *P. glaucophyllum* J.J. Smith. This species was first found by B.J.C. Verhey in 1897 near Turen, Malang, East Java. He presented the plant to Bogor Botanic Garden, where it flowered in 1899 and was described the following year by the famous Dutch botanist J.J. Smith.

The percentage duplication in orchid collections is 29.11% (Bogor and Purwodadi Gardens) and the smallest is between Purwodadi and Cibodas Garden (16.97%). These figures demonstrate that there is little duplication of effort for conservation of orchids between each of the gardens which, largely have unique collections. After the reorganization in LIPI (Indonesian Institute of Sciences) in 1987, the future development of collections has been more realistic so that each garden contains plants from its own most habitat and climatic type. Bogor collections are from the lowland humid plots; Cibodas collections are from the mountain of western Indonesia; Purwodadi has lowland dry plants and the Bali collection contains mountain plants of eastern Indonesia. Riswan and Sukendar (1988) proposed that *ex situ* orchid conservation should be undertaken not only in the botanic gardens, but also in the local gardens (Kebun Raja) in each province that are run by the provincial goverment, each one conserving local wild plants. In this way duplication between all the provinces of Indonesia, can be avoided, except for species that have a wider distribution. This strategy will ensure that many wild species will be saved from extinction, in particular endemic species and species with economic value, such as those of the genera *Vanda*, *Dendrobium*, *Arachnis*, *Phalaenopsis* and *Paphiopedilum*.

Bogor Garden and orchid knowledge in the Malesian region

The Orchidaceae is the largest and the most diverse plant family in the world. In nature there are probably more than 25,000 different species, native to almost every part of the world, growing at elevation from sea level up to 3,000 metres (de Vogel, 1988). In Indonesia, orchids form one of the largest plant groups with about 5,000 native species. According to Hawkes (1961), Irian (west and eastern part of New Guinea), has the richest orchid flora in the world, followed by Colombia, Brazil, Costa Rica, Borneo, Java and India. As far as orchids are concerned, the islands in Indonesia (Irian Jaya, Borneo, Sumatra and Java) are the richest in the Malesian region.

The history of orchid collections at these gardens, particularly at Bogor Botanic Garden is very closely linked with advances in orchid knowledge in the Malesian region. Most of the scientific studies on orchids in Indonesia have been concerned with their botany. Work began in this region in 1825 in Java, with the appearance of Bijdragen tot De Flora van Nederlandsch Indie (Blume, 1825), in which 110 genera and almost 300 species of orchids were described. Blume was the Director of Bogor Garden between 1822 and 1826. He was a diligent collector and hard worker and during his time the Garden vastly improved the number and range of its collections.

The most important contributor to knowledge of orchids in the Malesian region was J.J Smith. He was a member of the research staff of the Botanic Garden and Herbarium Bogoriense and had access to all material collected on expeditions throughout Indonesia. In 1905 he finished "Die Orchideen von Java", a work containing keys and comprehensive descriptions of all genera and species from Java and elaborate descriptions. This work was followed by "Die Orchideen von Java Figurean-Atlas" (Smith, 1911) with illustrations of all species and analysis of their flowers. He continued working on Javanese orchids, until his last major work was the text on the orchids of Java, which first appeared in Dutch in the mimeographed emergency edition of the "Flora of Java" by Backer, and later as an English version in "Flora of Java" (Backer & Bakhuizen van Den Brink, 1968). Other significant publications by Smith were "Die Orchideen von Ambon" (1905) and "Orchidaceae Novae Malayenses" (1908). Because of his important work and dedication to the garden, a monument was built in the Bogor Botanic Garden in his memory.

In physiology and genetic of orchids, Treub (1879) and Burgeff (1909) published articles on orchid embryogenesis and symbiosis between fungus and orchid. Both works prepared at the Bogor Garden's laboratory. After that time hardly any extensive work on orchids was undertaken in Indonesia. For general readers, however, Latif's books (1960, 1960 & 1968) and Sunarko's are worth mentioning. Comber who has collected plants for the Bogor and Purwodadi Gardens published a book on "Wayside orchids of Southeast Asia" (1981). He is also preparing a book on "Orchids of Java" (de Vogel, 1988). A large manuscript has already been produced and is being regularly revised. It contains more than 730 species with keys to genera and species and as well short descriptions and notes on the ecology and distribution of each. Articles by members of the Bogor staff have been delivered at many meetings and published in the Garden's journal, "Bulletin Kebun Raya". Worth mentioning are those by Irawati (1979 & 1983) on "Wild orchids in Bogor Botanic Garden" (Anggrek liar di Kebun Raya Bogor) and "Notes of *Dendrobium* collection in Bogor Botanic Garden" (Tinjauan pada koleksi marga *Dendrobium* di Kebun Raya Bogor).

Other activities of the Bogor Botanic Gardens have included the organization of special training sessions for the public in practical and technical aspects of orchid cultivation and conservation. These sessions have been organized 13 times in the last 10 years.

The future of the botanic garden and its orchid collection

Historically, the Gardens were founded with the dual purposes of furthering botanical research and experimentation on important crops. Introduced crop species, such as rubber, oil palm, cacao and coffee have now become the most important commodities for export in Indonesia and other South east Asian countries. The oldest oil palm tree in South east Asia, planted in 1844, is still alive in Bogor Botanic Garden.

Botanical research activities are concentrated on the evaluation of the biological characters for determining the taxonomic status, economic value and evaluation of conservation strategy. The researh was carry out together with the Treub laboratory and Herbarium Bogoriense (Botanical Division) and Zoological Museum (Zoological Division), Center for Research and Development in Biology (Puslitbang Biologi), LIPI. For examples, the Gardens and Herbarium Bogoriense are responsible for plant explorations and collections.

Other functions of Botanic Gardens are in public education, as a place for recreation and information services on landscape gardening, floriculture and also make network co-operation with botanic gardens in other countries for their survival.

The Bogor Botanic Gardens acts as a place for *ex situ* conservation of tropical plants in particular (including orchids), as complementary to *in situ* conservation. *In situ* conservation is managed under the Directorate General of Forest Protection and Nature Conservation, which is also the CITES Management authority. The Botanic Gardens are the CITES Scientific authority.

In conclusion, the Bogor Botanic Garden and its associated gardens have very important rôles, not only for the *ex situ* conservation for orchids, but also for other tasks such as for botanical research, education, recreation and information services. The problems we face in the future will not be easy to overcome as the Gardens are expensive to maintain. As Indonesia is a developing country any money spent must produce something of direct use to the people.

References

Backer, C.A. and R.C.B.v.d. Brink (1968). *Flora of Java*, vol. 3. Wolters-Noordhoft, Groningen, Nederlands.

Burgeff, H. (1909). *Die Wurzelpilze Der Orchideen Aus Samen*, pp 62. Gustav Fischer Verlag, Jena.

Comber, J.B. (1981). *Wayside orchids of Southeast Asia*. Heineman Asia, Singapore.

de Vogel, E.F. (1988). *Southeast Asean wild orchid : present knowledge, future nuclei of attention*. Paper presented on the Seminar of the 7th Asean Orchid Congress. Jakarta (in press).

Hawkes, A.D. (1961). *Orchids, their botany and culture*. Harper & Row, New York and Evanston.

Irawati (1979). Anggrek liar Di Kebun Raya Bogor. *Bull. Kebun Raya*, 4(1), 9-23

Irawati (1983). Tinjauan pada koleksi marga *Dendrobium*. *Bull. Kebun Raya*, 6(3), 53-57.

Latif, S.M. (1960). *Bunga anggrek permata belantara Indonesia*. Sumur Bandung, Bandung.

Latif, S.M. (1960). *Kembang nggrek I*. N.V Masa Baru, Bandung.

Latif, S.M. (1968). *Nomenklatur anggrek merangkap siklopedia*. P.T Saksama, Jakarta.

Riswan, S. & Sukendar (1988). *Anggrek dan permasalahannya dengan CITES*. Paper presented on One Day Discussion against The 7th Conference of CITES Member Countries in 1989. Puslitbang Biologi, LIPI, Bogor.

Smith, J.J. (1905). *Die orchideen von Ambon*. Landsdrukkerij, Batavia.

Smith, J.J. (1907). *Die orchideen von Java*. Erster nachtrag, Mit 2 Tafeln G. Kolff & Co., Batavia.

Smith, J.J. (1908). *Die orchideen von Java* Figuren-Atlas Ersters-Drittes Heft. E.J Brill, Leiden.

Smith, J.J. (1911). *Die orchideen von Java, Figuren-Atlas* Vierties Heft. E.J Brill, Leiden.

Soenarko, R. (1961). *Anggrek*. Penerbit Jambatan, Jakarta.

Treub, M. (1879). *Notes Sur L'Embryogenie de Quelques Orchidees*. Johannes Muller, Amsterdam.

Resumen

El Jardín Botánico de Bogor (Kebun Raya Bogor) fué fundado el 18 de mayo de 1817 por el Dr C.G.J. Reinwardt. Anteriormente conocido como S'Land Plantentuin Buitenzorg, o también como Hortus Botanicus Bogoriensis, este jardín es uno de los más antiguos de Indonesia, y los tres jardines asociados a él fueron establecidos con posterioridad: Cibodas (1874), Purwodadi (1941), y finalmente Bali (1959). En este articulo se discute la historia del jardín botánico como lugar donde se realiza la conservación de orquídeas *ex situ*. Se hace referencia también a la fluctuación de estas colecciones, así como al papel que desempeña el jardín botánico en el conocimiento de orquídeas, y en los problemas relacionados con su futuro desarrollo y mantenimiento.

The conservation of orchids in Sabah (Malaysian Borneo).

A. Lamb

Fruit Tree Division and Tenom Orchid Centre
Agricultural Research Station
Tenom, Sabah, Malaysia

Summary

A brief account is given of the current position in the Malaysian State of Sabah with relation to the conservation of habitats of wild orchid species. Of a probable 2,000 species of orchids in Borneo 1,500 are estimated to be found in the State. Many species of both horticultural and botanical interest are now considered by the author to be endangered or very rare in the wild due to several factors.

The preservation of orchid species is now carried out in a few centers in Sabah and good progress has been made with this programme. A programme for propagation of the most endangered species has begun in cooperation with several Botanic Gardens and one re-introduction project is being carried out.

The exploitation and stripping of rare orchids of commercial value from the wild has become a considerable concern in recent years, and to date, CITES has not been effective in stopping this trade. However, some measures have now been implemented to try and remedy this.

Orchids are practically the only plant group apart from *Rafflesia* species for which special conservation efforts have been made in Sabah and it is concluded that the success of the efforts made for this family will lead to other endangered plant groups being given similar treatment.

TROPICAL BOTANIC GARDENS: THEIR ROLE IN CONVERVATION AND DEVELOPMENT
ISBN 0–12–346850–7

Introduction

The State of Sabah forms the northern end of the island of Borneo and is politically one of the two states of Eastern Malaysia. It lies 5 to 7° north of the equator in the equatorial rainforest belt, with two monsoon seasons, the S.W. and the N.E. bringing heavy rains in May and June, and November to February respectively. There is a short dry season in most districts of the State from March to April. Average rainfall ranges from 2,000 mm. to 3,000 mm. per year in most districts with some exceptions such as Mt Kinabalu which receives more than 400 mm. per year. Temperatures in the lowlands range from a day-time average of 31 to 32°C, to 24°C at night. A 1°C drop for every 100 m. rise in altitude is the general rule for mountain areas.

Most of Sabah was raised above the sea in comparatively recent geological time and consists mainly of sedimentary rock formations with outcrops of limestone. However, other outcrops of volcanic soils, the granitic plug that forms Mt Kinabalu, and ultrabasic (ultramafic) serpentine hills and mountains have given rise to a considerable range of different soils, that has led to a very rich diversity of plant life.

75% of Sabah is made up of hills and mountains, giving rise to a very great range of altitudinal habitats. Of the land area, the narrow western coastal plain and the interior valleys have long been developed for agriculture. Much of the hilly interior is Commercial Forest Rcserve. Other areas are under shifting cultivation. The bulk of the large scale agriculture at present (and for future plans) is in the alluvial river plains and on the fertile peninsulas of the east coast. Fairly large coastal mangrove and tidal swamp forests still exist there. It should be pointed out that several types of rainforest containing wild orchids cover the land area. The coastal nipa palm and mangrove forests are generally very poor in orchids species, but the coastal swamp forests that usually occur behind them can be quite rich in epiphytic orchids. Raised coastal podsolic terraces with heath or "kerengas" forest also contain quite a number of orchid. The lower river valleys and low hills were generally covered in the majestic Dipterocarp forests, but are now nearly all logged or cleared for agriculture. Some remain in Commercial Forest Reserves or as remnant patches. Here though, terrestrial orchids are few. There is quite a diversity of epiphytic species in the canopy but this can only be seen along the more open river banks or where the forest has been freshly felled. The Dipterocarp forests reach to about 750 m. (2,500 ft) altitude but, within this zone, several other distinct forest types occur. Plateaux with "kerengas" or heath forest at 500 m. (1,500 ft) or above, show a lot of species replacement when compared to the coastal heath forest. These are some of the richest habitats for orchids both in diversity and number of plants. Where limestone or ultrabasic soils occur there is quite a change in species composition both in trees and orchids and such areas usually contain a unique and diverse orchid flora. Above 760 m. the hill dipterocarp and lower montane oak/chestnut forests start, rich in both terrestrial and epiphytic orchids. As these forests merge into moss or cloud forest, around 1,300 m. (4,000 ft), orchids tend to become even more plentiful, especially along rivers or on ridge tops or where exposed cliffs appear. Above 1,800 m. (6,000 ft) orchid species diversity drops but in some areas they form a considerable proportion of the total number of plants per unit area. For example, at 3,000 m. on Mt Kinabalu, the forest floor under *Leptospermum* and Ericaceous shrub can be a solid mass of orchids, but with less than 50 species, dropping to less than 20 species at 3,300 m. (11,000 ft).

Of the 76,115 sq. km. (7.6 million ha.) of Sabah, only 18,000 to 20,000 sq. km. (1.8 - 2 million ha.) or 25% has been assessed as suitable for agricultural development. Of this, less than 500,000 ha. have been properly developed for agriculture at present. However, large areas of existing shifting cultivation are carried out on what are considered to be unsuitable steep slopes. Sabah has a population of 1.3 million with the highest annual growth rate in Malaysia of 3.6%. This will mean a rapid increase in pressure for more land for agricultural development, since 70% of the population lives in rural areas.

Over 50% of land in Sabah falls into various categories of forest, water catchment, or special reserves. The majority of these are commercial forests and it is estimated that within the next decade all the primary forest in these areas will have been logged or re-logged for a second or third time. National and State Parks that provide more or less complete protection only cover 2.6% of the land area. To this can be added special Wildlife Reserves in which further habitat destruction is unlikely under present circumstances. The Sabah Foundation with 3,000 sq. km. of forest concessions, have set aside two large areas as conservation areas which are not suitable for large scale timber extraction. This could bring the total area of protected land in Sabah to nearly 5%.

The Orchid Flora

It was estimated by Dr T.P. Whitmore (1975) that the Malaysian region contained 3,000 - 4,000 species of orchids, in over 200 genera, of which 40% are endemic, with even higher endemism for the species.

Unlike surrounding countries such as the Philippines, Java, Malaya and Thailand where fairly comprehensive accounts of the orchid floras exist, Borneo has little published material; some literature published before the war is now out of print. The last enumeration of orchids was by Masamune in 1942, but in recent years new records and many new species have been described for Borneo. However, it also appears that many species in the region have been described under several different names and hence a confused picture of the orchid flora of Borneo has existed for many years.

In 1983 a group of orchidologists got together and with the help of several herbaria, notably that at the Royal Botanic Gardens, Kew, started the Orchid Flora of Borneo project with the Kew Herbarium playing a central rôle. In the last few years Kew have also been preparing an updated check-list of orchids of Borneo.

From all the information gathered so far, this author estimates that Borneo has probably 2,000 species of orchids in nearly 140 genera, of which nearly 1,500 species can be found in Sabah. Recent work by J.J. Vermeulen on the genus *Bulbophyllum* indicated that of nearly 200 species, almost 20%, were unnamed. This author has found a similar situation exists for the sarcanthine orchids.

It would appear that of these species at least half are found in the lowlands below 760 m. (2,500 ft) and the rest in the highlands. the 4,101 m. high Mt Kinabalu (13,455 ft), probably boasts a total of nearly 1,200 species, including those found on its lower slopes which now largely fall outside the State Park created around it.

Following the publication of "Classification of Orchids" (Dressler, 1981), all six subfamilies are represented on Mt Kinabalu, making it a fascinating mountain on which to study the evolutionary aspects of the orchids represented, from the most primitive to the more recently evolved.

Of the various Orchid Centres set up in Borneo, the Tenom Orchid Centre which is covered in some detail in this paper contains a living collection of over 453 species in 104 genera including several species from Sarawak, Kalimantan and Brunei, but largely representing lowlands species of Sabah. From the author's observations the centre has probably collected most of the lowlands species of horticultural interest either for breeding, cut flowers or as ornamentals and pot plants. From surveys, field trips, and a good knowledge of the habitats in which these species grow; each of these orchids has been given a conservation rating, following the IUCN categories of endangerment.

Of the species in the Tenom collection, 28 appear to be endangered, 39 vulnerable, and 58 are rare. Of the horticultural species, 19 are endangered, 18 vulnerable and 10 rare. These figures will naturally grow when the status of other species in Borneo is determined. these figures would appear to be alarming, but the situation is not quite as bad as it seems if we compare the conservation of habitats and the preservation of the species in various collections, including botanic gardens around the world.

Long-term conservation of orchid species in their habitats

It is clear that there is a rapidly increasing number of orchid and other plant species under threat by forest destruction and loss of habitat in Sabah. Due to the rapidly increasing rural population, it can be assumed that in the future all land suitable for agriculture will eventually be developed. Existing areas of land already developed outside this area by shifting cultivators should stabilize as the Agriculture Department and State Government persuade these rural farmers to adopt a more permanent system, since they presently exist at or below poverty level.

Since commercial logging provides the State Government with a very considerable percentage of its revenue, it is now clear that this logging will continue in these commercial forest reserves and that all primary forest within these reserves will be logged within the next five years at the present rate. The pressure to relog areas will also increase and this is seen as having very severe consequences on the future regeneration of the Dipterocarp forests.

These areas have now been threatened by a further, more destructive factor: changing climate and the rise in temperature producing more severe droughts has made forest fires a real future threat. The huge ecological disaster in Kalimantan in 1983 when a six months drought allowed huge forest fires and a vast area of forest was destroyed, was widely publicised internationally. Not so well publicised was that 20% of the area of Sabah was burnt in similar fires. Aerial and other surveys showed that over 80% of the area burnt was in logged forest. Such fires probably caused a complete loss of orchid species in those areas, apart from in small isolated patches of forest not so severely burnt.

These fires not only added vast amounts of carbon dioxide to the atmosphere, but the land area affected subsequently reflects more of the sun's energy. What is not so well understood by the public is that the biomass of a virgin tropical forest absorbs over 70% of the sun's energy. When this biomass is reduced (as calculated) by up to 60% by commercial logging, or even more, with forest fires, the extra energy not absorbed and reflected is considerable, not to mention the reduced carbon dioxide uptake and oxygen production by such a forest biomass.

As the timber from the Dipterocarp forests diminishes there will be very great pressure to move higher up the mountains to extract *Agathis* and other hill species, even as high as 1,000 m. and above.

If we accept that very little is going to change this situation, it becomes increasingly urgent to ensure better protection for what other areas are left outside commercial forest reserves. Illegal logging along the edges of the State and National Parks has been taking place and this has to be publicised and stopped. If given full protection the areas left could conserve a very great proportion of the orchid flora.

The main areas under State Parks include some islands with largely undisturbed coastal forests. These are Pulau Gaya and Pulau Tiga. The former has about 2,000 ha. of forest and probably contains about 50 species of orchids. It is the most important and could be the best locality to conserve the seashore species *Arachnis hookeriana*, a very valuable parent for many hybrids, which has now disappeared from the coast of Malaya and is known from only a few sites in Sabah.

The Tawau Hills Park on the east cost of Sabah contains 27,972 ha. and has been partly logged, but it contains some lowland forest and extinct volcanoes with mountains rising to 1,300 m. A survey of the orchid flora has not been carried out, but it contains some important *Phalaenopsis* species and is the only park on the east coast.

The Crocker Range National Park, Sabah's only National Park, in size nearly 140,000 ha., is an important water catchment area for a large proportion of the population of Sabah, with protected hills from 300 m.

to a continuous line of peaks over 1,300 m. With considerable areas of moss forest, it is rich in orchid species found in the other unprotected hill forest areas.

The Mount Kinabalu State Park, with its 4,101 m. granitic massif lies in an area of 75,350 ha. It has many forest types in various vegetation zones at different altitudes and with a variety of different soils, adding to the diversity of vegetation with unique floras occurring on the ultrabasic soils in particular. The Mt Tamboyukon (2,640 m.) is wholly ultrabasic and falls within this park.

Although the lower slopes of Mount Kinabalu have been cleared along much of the boundary some areas of lowland forest are found at 400 - 500 m.

With an estimated 1,200 orchid species found in these forests it becomes the most important conservation area. The difficult terrain and long boundaries make it impossible for the park authorities to stop the poaching of species, and though the habitats are safe, rare, sought after species have become endangered in recent years. 5 species of *Paphiopedilum*, especially *P. rothschildianum* and *P. dayanum* are in the most danger. Species of *Phalenopsis*, *Renanthera*, *Dendrobium*, *Paraphalaenopsis*, *Arachnis and Dimorphorchis* are also threatened by this poaching.

The Maliau Basin, a conservation area of nearly 300 sq. km. under the control of the Sabah Foundation contains many rare and newly recorded species, and, together with the lowland forest Study Centre in the Danum Valley with areas of lowland ultrabasic forest, it constitutes a major addition to habitat conservation. The Sabah Foundation, in addition to its research studies in the Danum Valley, has initiated Nature Clubs in schools throughout Sabah to make the next generation more aware of the wealth of the forest resources and encourages these clubs to visit the area.

The newly created Tabin and Kalumba Wildlife Reserves indicate that they too contain a diverse orchid flora that should remain intact, and the latter includes some very interesting coastal swamp forest.

There remain, however, several unique ultrabasic areas, kerengas forests at different altitudes as well as coastal forest areas that need to be conserved. If Sri Lanka, which is smaller in area than Sabah, can conserve over 10% of its land in Game and other Reserves, despite the pressure of a population of over 15 million, it would appear that Sabah could easily also plan for this also. IUCN and WWF-Malaysia will continue their efforts, it is hoped, to press for more areas of rain forest to be fully protected.

Overall if these areas mentioned can be fully protected against logging and other development the long term conservation of Sabah's orchid flora is very good.

There remains a problem, however, with the smuggling of rare horticultural species from these areas. CITES has not prevented these plants from getting into Europe, the U.S.A. and Japan where there is a large demand for them. In Sabah increased efforts are being made to catch both the local middlemen and the collectors. In Europe also, court cases are impending against some of these dealers. Though the number of plants brought overseas has been reported to be quite considerable, it is felt that that the principles behind CITES must stay and that it is its implementation that needs to be improved. Furthermore the recent incidents involving orchids should not panic the authorities into moving the species involved onto Appendix I of CITES for which no trade is allowed. This could be very detrimental to the propagation of these species from seed or by meristem culture and their subsequent distribution to commercial nurseries and breeders. Once these species can be made commonly available through artificial propagation then their commercial value will drop and collecting from the wild would be greatly diminished.

In fact, in Sabah, seedlings of *Paphiopedilum rothschildianum* have been brought back from Europe for a re-introduction trial into the habitat. This is now being monitored by the State Parks. A similar offer has now come through the American Orchid Society to try the same method for the endangered Sarawak species, *P. sanderianum.*

Short-term conservation – preservation of orchid species in collections and botanic gardens.

With the rapid loss of habitats it is now clear that the only way to safeguard many species is to collect survivors from the wild and maintain them in collections and botanic gardens, and where possible to spread the risk between several institutions. At a later date certain species can then be re-introduced into areas where exploitation of the habitat has ceased and some form of protection is provided by making it a reserve.

This form of conservation has been around for some time in Sabah and started with a small collection of lowland species at the Forest Research Centre, Sepilok, (c. 100 m a.s.l.) on the east coast in 1977. This was followed by the setting up of the Tenom Orchid Centre (c.200 m. a.s.l.) in the interior, in an area which was at one time rich in species, but where agricultural development was causing a rapid loss of habitat. This was set up and managed by the Agricultural Department, for the Ministry of Environment. In 1986 it was decided to set up a similar centre at Poring Hot Springs (c. 550 m. a.s.l.) in the Kinabalu Park, to attract more tourism to the park, with a duplicate set of all the species from Tenom. For certain species, the higher rainfall and cooler temperatures at Poring have proved ideal. The Poring Centre is set in a partly forested area of nearly 4 ha. and is managed by the Sabah Parks. Therefore, for lowland species, the risk of losing plants to disease or other factors is spread over three centres covering a range of climatic habitats.

At present with less threat to mountain species, only the Mountain Garden at the Kinabalu Park at 1,550 m. (5,100 ft) contains a limited number of orchid species and is restricted to those from Kinabalu alone. However, a new centre is to be established this year (1989) at the same altitude in the Crocker Range in the Tambunan District. It is hoped that orchids from mountains all over Sabah can be established here, even if they are not endangered, as this will facilitate the enumeration of species for the Orchid Flora of Borneo Project. The Crocker Range Centre will also contain species of *Rhododendron* and *Nepenthes*, as does the Mountain Garden in the Kinabalu Park.

At present there is no botanic garden in Sabah, Sarawak or Brunei, but the Sabah State Parks have been giving serious thought to establishing such a garden at the Poring Hot Springs at the foot of Mt Kinabalu.

The efforts in Sabah to preserve its orchid species have been complimented by the setting up of a similar Orchid Centre, by the Forestry Department, Sarawak, near Kuching. In Peninsular Malaysia, the Malaysian Agricultural Research and Development Institute (MARDI) have a research unit for orchids, including a propagation facility. They have established two centres for the collection of wild species, at Serdang in the lowlands, and also in the Cameron Highlands. The Federal Government of Malaysia has also set up a National Orchid Centre in the Lake Gardens in Kuala Lumpur.

This paper will highlight the work of the Tenom Orchid Centre in more detail.

The rôle and objectives of the Tenom Orchid Centre and its establishment.

The establishment of the Centre was begun in September 1979 and it was located within the Agricultural Research Station, Tenom at 200 m. a.s.l. It is contained within an area of just under one hectare with a small stream running through it and it has access to roads and an irrigation canal. The facilities include a small store and nursery area, staff quarters for a supervisor and two daily-rated employees and a network of permanent concrete paths giving access to the plant houses with different shade regimes. The rest of the area is landscaped with trees that are suitable for establishing wild orchid species. By the end of 1984 these facilities had been increased with the construction of a small building to accommodate visiting research staff or orchidologists, with a small laboratory/office and a small herbarium facility.

Up to 1986, the setting up of the Centre and its operation was carried out by the Department of Agriculture, financed and greatly encouraged by the Ministry of Environment. When this Ministry changed its rôle to mainly promoting tourism, the Centre was taken over by the Ministry of Agriculture. The rôle and objectives of the Centre have now been updated with a programme of work clearly defined to achieve its objectives.

The collection and cultivation of wild species.

The first priority is to collect lowland species of wild orchids particularly those of horticultural value or those that are thought to be endangered through the loss of habitats destroyed in the course of land development and logging. Lowland species are considered by us to be those found up to 760 m. (2,500 ft) which have been found to grow well and flower even at 200 m. This altitude generally corresponds to the main loss of lowland forest through shifting cultivation along populated valleys. Even species that occur at 1,000 m. a.s.l. have been found to grow at Tenom but they rarely flower.

The second priority is to study the conditions of shade, substrate and humidity under which the species thrive best, so that survival of the species in the Centre is more assured, and later to document these conditions precisely.

The third priority is to propagate or collect a reasonable population of plants of those species thought to be endangered by loss of habitat or commercial collecting.

The above programme entails field trips that survey areas for species in need of collecting before the forest goes, surveys to establish areas with a rich flora of orchids as well as actual collecting. Such field trips also entail the collection of soil and rocks and leaf litter for terrestrial orchids to establish those orchids that grow on different soils.

Generally, for the terrestrial orchids, three habitats rich in species have been determined as of particular interest. These are, firstly, podsolic leached sandy soils where the lower crown of scrub forest (kerengas) allows more light to penetrate the forest floor, and where sometimes there exists a thick peaty leaf litter. Secondly, but very restricted in occurrence in Sabah, are limestone areas, particularly the rocky areas where vegetation is more sparse, and "rendzina" type soils rich in humus develop on ledges or crevices. The third type where it occurs in the lowlands are the ultrabasic soils, particularly rich in unusual species that grow on them. There is also a necessity for the constant replenishment of the correct leaf litter. Leaf litter from podsols rich in tannins and very acidic does not suit, for example, the orchids of the limestone areas.

The other main soil groups most commonly found are usually associated with large dipterocarp forest trees providing heavy shade and competition. Hence terrestrial orchids are scattered under these conditions, although a diverse range is found, particularly where ridges give exposure and allow more light in.

There soils are derived from the mixture of sandstones, mudstones and shales or from the alluvial soils in valleys. Various other types of soil occur but the Centre plans to cater for these five main groups.

For epiphytic orchids, the type of tree they grow on is also of some importance. Different shade regimes of approximately 70%, 50% and 30% shade have been developed in the plant houses. We have standardised with three types of branch or tree trunk supports for orchids that are cut and arranged as branches attached to steel bars coming from reinforced concrete support pillars and are favoured by orchids. The ones we choose do not rot quickly and hopefully last for up to five years.

These are "kulinopapa" (*Vitex pubescens*) and tree-ferns (*Cyathea* spp.). For upright trees with branches that are fitted and attached to concrete bases, "ironwood" or "belian" (*Eusideroxylon zwageri*) is used but its weight and availability pose limitations on it being obtained.

This programme is further increased by surplus plants of wild species in which we have an interest in growing in greater numbers, being established on living trees. Three indigenous wild species are favoured – "yellow flame" (*Peltophorum ferrugineum*), "queen flower" (*Lagerstroemia speciosa*) and "kulinpapa" (*Vitex pubescens*) and, of lesser importance, *Casuarina nobilis*. Two introduced species, namely the Indian cork tree (*Millingtonia hortensis*) and especially, *Gliricidia sapiens* are used. Another unidentified species growing in the stream at the Centre has also proved very good for the purpose.

As orchid plants are collected and established in the Centre they are identified if possible, numbered and planted out at a place in the Centre where conditions are nearest to those found where it grows in the wild, depending on shade regime and humidity as well as the soil type when that is relevant.

The numbers, name of the species, locality data and other information is then entered on an accession card and into a Store ledger and indexed as to its location in the garden, according to a locality plan. This accessioning was only started in 1986 and is still not completed. For each species either at the time of collection or at a later date, herbarium material and a spirit collection of the flowers is made.

In future this will enable up-to-date print outs to be made of all the existing species found in the Centre, together with any detailed information on the plants. This will be stored in the computer, based at the Research Station. Plants dying can be recorded and written off so that accurate records can be kept. The accession lists can then be circulated to botanic gardens that are interested.

The Orchid Flora of Borneo project

Another rôle of the Centre is in species identification, since without a knowledge of what species occur in Sabah we will not know what we need to conserve. To this end the Centre has set aside funds and provides facilities to encourage and assist orchidologists working on the "Orchid Flora of Borneo". This project was started in 1986 and with the availability of accommodation, work space, stereo microscope with camera lucida, and a limited herbarium facility, local orchidologists are now able to use the Centre as a base for drawing and describing the numerous lowland species growing at the Centre. With an estimated 2,000 species in Borneo (with perhaps up to 50% in the lowlands) it is expected to take 20 years or more to complete. It is planned to be published in volumes of 100 species, with Volume 1 appearing in 1989. At present a further 6 volumes are being prepared by local and overseas orchidologists who are "experts" in their field. Sabahan orchidologists involved in this project are Encik Chan Chew Lun working on *Coelogyne* and related genera and Encik William Wong on *Dendrobium* and related genera. Foreign orchidologists who are preparing volumes include the author working on sarcanthine and other orchids, J.J. Vermeulen on the genus *Bulbophyllum*, and taxonomists at the herbarium at the Royal Botanic Gardens, Kew (J.J. Wood and Dr P.J. Cribb) and the Rijksherbarium at Leiden in Holland (Dr E. de Vogel). Further help is provided by taxonomists Dr Seidenfaden in Denmark and Dr Eric Christenson of the Marie Selby Gardens, U.S.A. Therefore, the major task is a co-operative effort between local orchidologists, the Tenom Orchid Centre and the Royal Botanic Gardens, Kew, England, which assists with identifications, the maintenance of herbarium material, and research into other data on each species, with the Sabah Society agreeing to publish the first volume.

Other organisations that are providing considerable help are the Singapore Botanic Gardens, the Sabah Parks, the Sabah Forest Research Centre, the Forest Department and National Parks, Sarawak, the Rijksherbarium, Leiden, Holland and several botanists in other institutions. It is hoped that the Herbaria at Bogor, Indonesia and Forest Department Brunei, as well as the Botany Department of the University Kebangsaan, Malaysia branch campus in Kota Kinabalu will become involved in these efforts and provide future authors. A visiting professor, Dr John Beaman, has provided considerable help with collections in the past for Volume I, and the present taxonomist, Kamaruddin Mat Salleh is considering involvement.

However, the Orchid Herbarium at Kew Gardens have provided the main back-up since they have the largest collection of herbarium material and literature with which to help with identification and nomenclature and have subsequently made assistance for the Orchid Flora of Borneo project one of their official projects also. The staff of this herbarium have made field trips to Borneo and offered advice and help with regard to herbarium procedures, and instructions on how to document living collections of wild orchid species.

Preservation of wild orchid species.

Andree Miller in both the Third and the Fourth ASEAN Orchid Conferences detailed the conservation problem in Papua New Guinea and concluded that preservation in the present circumstances was the best path. A similar pattern would appear to be true in Malaysia in that it is very hard to conserve species in their habitat, until education and greater awareness of the botanical heritage that needs conservation becomes accepted politically as well as generally by the public, and that planning of natural resources finally defines areas to be fully protected.

The collecting of wild species and preserving them in a living collection is therefore the main rôle of the Centre. However, to this must be added the release and distribution of seeds or seedlings of species of orchids of horticultural interest to orchid breeders, other institutions and growers. For the rarer species, at present the greatest danger is from collectors, either private or commercial who are stripping desired plants, and from agricultural and forestry development and land clearing for other purposes. For these their preservation is urgent and is best followed by the production and distribution of seeds and seedlings to reduce pressure from collectors in particular.

For most commercial breeders and collectors, it is a very expensive business to make a special journey to Sabah to seek out, sometimes in vain, and collect the desired species. Hence the market value for them is high, and this encourages local collectors to provide the plants. Furthermore, local collectors will strip the more beautiful species of orchids for sale in the local market, the less plentiful, the higher the price. This is illustrated by the stripping of the more widespread and common but locally favoured *Phalaenopsis amabilis* and the more specific stripping of the rarer and highly valued *Phalaenopsis gigantea*.

Another rôle for the Centre has therefore been set and that is the propagation of endangered horticultural species, and species of botanical interest. Vegetative propagation is possible with a few species but it is slow and the best avenue is crossing plants of the same species to produce seed pods or by tissue culturing a plant. Facilities for flask and tissue culture are not available, nor are they planned for at the Centre. Instead, provisional arrangements have been made with Singapore Botanic Gardens and the Kew Gardens micropropagation unit to germinate seed pods, and also in the past with the National Botanic Gardens for Australia at Canberra. However, this project has made a slow start.

A second project under this programme is to send seed pods to seedbanks, for further distribution to growers who want the species. The Royal Botanic Gardens, Kew has established such a seedbank, whilst the Australian Orchid Foundation is also successfully operating one. A visit of researchers from the Sydney Botanic Gardens have suggested that the Centre considers sending seed pods to this foundation and this is being planned for. The programme for seed production is therefore to be given priority in the coming years.

Research by M. Clements of the Australia National Botanic Gardens in Canberra, on the mycorrhiza of different orchid species, which are essential for the germination and growth of orchid seedlings in the wild, will in future, greatly help reintroduction of species back into their wild habitats, particularly from those habitats previously stripped of plants by collectors.

In 1987 a first trial for re-introducing a rare species back into a potential forest habitat was tried in Kinabalu Park. Seed pods sent from Sabah of the endangered slipper orchid, *Paphiopedilum rothschildianum*, were successfully germinated and grown in West Germany by Dr Haas and Dr Grell of the Max-Planck Institute and Biophysics in Frankfurt. They visited Sabah to establish seedlings back into a habitat similar to that from which they had been completely stripped. This trial, if successful, will be of great importance to future conservation of other endangered species of plants (Grell et al., 1988). An offer through the AOS for *Paphiopedilum sanderianum* to be similarly introduced is also being considered.

With respect to tissue culture, the Centre has contact with the MARDI Orchid unit which has access to tissue culture facilities. However, they need to carry out basic research on cultures required for the different genera of horticultural interest. Although cultural requirements are known for genera such as *Vanda* and *Arachnis* this is not so for other genera and species.

For all these avenues of conservation there must be an information service. Through the Threatened Plants Unit of the World Conservation Monitoring Centre (WCMC), the Botanic Gardens Conservation Secretariat, and CITES, it is hoped that botanical institutes world-wide will make available lists of threatened plants and those that they have in cultivation. The Kew Orchid Herbarium have suggested that they might well do this for orchids specifically and in this respect they have started an Orchid Research Newsletter that distributes names of all researchers and institutes requiring various orchid species. As well as this, several countries have established orchid seedbanks and this appears to be one of the greatest avenues for the future preservation of orchid species. The Orchid Specialist Group of the Species Survival Commission of IUCN is also pressing for the preparation of check-lists of orchids of different countries and they also advise CITES on which species should be placed on the various appendixes under the convention. The Tenom Orchid Centre keeps the group updated on Borneo species.

The WCMC Threatened Plants Unit are compiling lists of all endangered plants world-wide, using a computer database. For Sabah and Borneo in general they are relying on data provided partly by the Tenom Orchid Centre.

The Kew Orchid Herbarium is also compiling data on computer on all orchid species found in Borneo and may keep records eventually of all living collections of orchid species world-wide, and through the Botanic Gardens Conservation Secretariat a free flow and exchange of rare species in need of preservation will be possible in the near future. Hopefully the Tenom Orchid Centre which has already started such an exchange with botanic gardens will be an integral part of this conservation scheme.

Preservation of early cultivated hybrids

Orchid breeding has advanced at an incredible pace world-wide with over 60,000 registered hybrids. The breeding of hybrids not only depends on the availability of wild species but also on some of the early hybrids that today are often being discarded and lost. Some of these early hybrids are still important to breeders.

For long term planning it is suggested that an orchid cut-flower industry, requiring the backing of breeding work, has potential in Sabah. To this end the Tenom Orchid Centre is collecting a limited number of orchid hybrids and, where possible, the older hybrids that are being discarded by many growers. Such a collection, together with the wild species, some of which have never been utilised as yet, but have great potential in producing new hybrids, will form an initial base on which a small, limited and specific breeding programme could be started.

The other rôles of the Tenom Orchid Centre

Apart from the above mentioned objectives, the Orchid Centre has provided a facility to popularise orchid cultivation locally and make people aware of the tremendous variety and beauty of their own local species, and the need and reasons for conservation.

In doing so there have been some bad side effects, particularly in respect to increasing awareness of some of the rarer and sought after species that are then stripped. However, this is inevitable when the basic approach used is that conservation or preservation will not progress without knowing the species that need such preservation.

Aside from visitors to the Centre, exhibitions designed to increase an awareness of the rich flora of orchid species have been displayed at orchid shows of the Sabah Orchid Society in which the Centre has actively participated, together with various publications on orchid species of Sabah in various journals.

Progress of the programmes at the Tenom Orchid Centre

Although many of the advances made already have been outlined earlier in this article, more detailed progress on the collection of wild species of interest is now added.

The wild species collection: the Centre has so far largely restricted collecting to nearby localities in the Tenom Valley, other areas in the interior as far north as Ranau, and along the west coast. However, the Centre avoids asking local people to collect for the Centre for obvious reasons.

From studies and observations to date it is roughly estimated that there must be at least 2,000 species in Borneo (10% of the orchids world-wide) of which 1,200 to 1,500 could be found on the slopes of Mt Kinabalu. Adding the rest of Sabah this could mean Sabah contains at least 1,700 species of which 40-50% are probably lowland species.

The priority of the Centre has been to collect the rarer orchids from the nearest accessible areas. A proper system of documentation of the collection has fallen behind the collecting itself but having now collected many of the rare species this should be rectified in the period from 1989 to 1990.

It is now estimated that over 450 species are growing in the Centre. For many, judging by their growth compared to those in the wild, optimum growing conditions have been provided. Many of the trees planted in the Centre are also now of sufficient size to provide a permanent base for many epiphytic species and these are being planted up with surplus plants of various species, so that there are now a total of well over 3,000 plants established.

Appendix 1 includes a list of species of more horticultural importance, together with species of botanical interest, particularly those endemic to Borneo that have established well. The list covers over 80% of the total number of species collected to date in 104 genera.

A limited number of species for which adequate plant stocks have been established at the Centre and for which an excess of material existed, were made available to various botanic gardens with living orchid collections, so as to spread the task of preserving species. Such institutions also should be in a position to produce seed pods in future, and distribute seedlings to growers. The Poring Orchid Centre will also act as an additional safety measure against species losses through disease and pests by having a duplicate collection of most of these species. The botanic gardens and other institutions supplied in the years between 1980 and 1988 were the Royal Botanic Gardens, Kew, Munich Botanic Gardens, the Arnold Arboretum, Singapore Botanic Gardens, Sydney Botanic Gardens, MARDI Orchid Unit; a few specific orchid species supplied to Zurich Botanic Gardens, Edinburgh Botanic Garden and Leiden University Botanic Garden.

In respect of the propagation and distribution of orchids through seed-pods, a small start was made in 1983 and 1984 with limited success, due to our inexperience in controlling fungi, harvesting pods at the correct time, and delays in posting the pods. Nevertheless, pods have been sent to Kew Micropropagation Unit, the Singapore Botanic Gardens and to the Australian National Botanic Gardens, Canberra.

Success has been reported for the rare *Dendrobium spectatissimum* and *Paraphalaenopsis labukensis* at Kew. Singapore Botanic Gardens have had good success in germinating large numbers of *Phalaenopsis gigantea* and *Benanthera matutina*, and one hybrid cross of interest also successfully germinated at Singapore has been *Porphyroglottis maxwelliae* X *Grammatophyllum speciosum* and hopefully also *Vanda dearei* X *Paraphalaenopsis denevei*. At Canberra success has been achieved with the endangered *Paphiopedilum rothschildianum*. In 1985 to 1986 and 1989, seed pods of the rare *Arachnis grandisepala* were distributed to Singapore and Kew Gardens.

In regard to the collection of species of more botanical interest, the following very distinctive Borneo species have been successfully established at the Centre - *Dimorphorchis rossii* and other varieties of it for which names have not been published, *Porphyroglottis maxwelliae, Grammatophyllum scriptum, G. speciosum, Vanilla pilifera, Paraphalaenopsis denevei, P. labukensis, P. laycockii, Vanda hastifera, Vanilla kinabaluensis, Arachnis longisepala and Bulbophyllum beccarii.* Many other species of botanical interest appear to be new and await description; among them several species of *Calanthe* and *Phaius*, several *Dendrobium* and *Trichoglottis* species, and many interesting Bulbophyllums. Lately a new variety *alba* has been discovered for *Phalaenopsis modesta*. Amongst the rare new species of horticultural interest well-established at the Centre are *Arachnis longisepala*, a newly described species, *Renanthera (Renantherella) auyongii, Renanthera bella, Calanthe* sp. and *Spongiola lohokii*; *Renanthera bella* was used for hybridizing, so far without success.

Orchid Flora of Borneo project: Over 400 botanical plates of different species have been completed, as well as their respective colour photographs. Volume 1 is expected to be published in 1989 with Volumes 2 to 4 following over the next three years, but that will depend upon the success of Volume 1. If the latter comes up to expectations, and is accepted, then closer co-operation with other orchidologists and institutes will be sought for subsequent volumes. However, this is a matter for the authors, as the Centre is only one of the centres being used and is there mainly as a facility to encourage further work on the flora.

Collections of hybrids: A good start has been made on this collection with plants being donated by the Singapore Botanic Gardens, other nurseries and the President and members of the Sabah Orchid Society. Several of the early *Vanda* and *Aranda* hybrids in particular have been collected but some difficulties in naming these hybrids is being experienced. In return, the Tenom Orchid Centre has helped the Sabah Orchid Society for many years with creating and setting up landscapes with orchids during their annual orchid show in Kota Kinabalu and at ASEAN shows.

Conclusions

The following conclusions can be drawn from this paper:

1. The need for orchid conservation in Sabah has been given due recognition by the State Government, and the public is now more aware of the rich diversity of the orchid flora in the State.

2. The conservation of habitats for montane species is very satisfactory, but for the lowlands a much greater effort is needed especially for species found in special forest types that are normally restricted in areas such as coastal and lowland podsol heath forests, and forest on lowland limestone and ultrabasic soils in particular.

3. The State should increase the Totally Protected Forest Areas from 5% - 10% in view of the large amount of land not suitable for agriculture, and this increase should be of lowland forests with particular emphasis on the unique forest types often rich in orchids.

4. It is concluded that the preservation of orchids that are rare or endangered in their habitats, and kept in the various orchid centres is quite satisfactory, but further efforts are needed now to propagate these plants to make them less valuable in the market-place, and so discourage further illegal stripping and trade.

5. Without a knowledge of the species that exist in Sabah thc needs for conservation cannot be fully realised. In this respect the Orchid Flora of Borneo project should continue to be given full support both from within and outside of the State.

6. It is concluded that CITES is essential legislation to restrict and stop the stripping and trade in endangered species, but its implementation in Sabah has not been organized. It is recommended that all orchid species be kept in Schedule 2 so that propagation of rare and endanger species on a large scale can easily be facilitated.

Acknowledgements

The author wishes to thank the Director of Agriculture, Sabah, for permission to present this paper.

References

Beaman, R.S. et al. (1985). Drought and forest fires in Sabah in 1983. *Sabah Society Journal*, 8, 10-30.

Dressler, R.L. (1981). *The Orchids - Natural History and Classification*. Harvard University Press.

Grell, E. et al. (1988). Re-introducing *Paphiopedilum rothschildianum* to Sabah, North Borneo. *American Orchid Society Bulletin*, 57(11), 1,238-1,246.

Masamune, G. (1942). In: *Enumeratio Phanerogamarum Bornearum*, 110-226.

Miller, A. (1982). Conservation or Preservation? In: *Proceedings of the 4th ASEAN Orchid Congress*, 87-92. Philippine Council for Agriculture and Resources Research and Development.

Vermeulen, J.J. (in prep.). *The Orchid Flora of Borneo* Volume II: *Bulbophyllum*.

Whitmore, T.C. (1975). *Tropical Rain Forests of the Far East*. Clarendon Press, Oxford.

Resumen

A continuación se presenta un resumen de la situación actual de la conservación de hábitats de especies de orquídeas silvestres en el estado malayo de Sabah. De un total de 2.000 especies de orquídeas presentes en la isla de Borneo, se estima que 1.500 especies se encuentran en Sabah. El autor considera que muchas especies de interés botánico y hortícola se encuentran en peligro, o son muy raras en estado silvestre debido a varios factores.

La conservación de las orquídeas es llevada a cabo en varios centros de Sabah, y este programa ha progresado mucho. Con la cooperación de varios jardines botánicos, se ha iniciado un programa de propagación de las especies que estan en más peligro de extinción, así como un proyecto de reintroducción de las mismas.

En los últimos años ha existido una preocupación por la explotación y el despojo de su medio natural de orquídeas raras de interés comercial, pero hasta la fecha, CITES no ha podido detener, de forma eficaz, dicho comercio. De todas maneras, se han puesto en efectivo nuevas medidas para tratar de remediar este problema.

Las orquídeas son practicamente el único grupo vegetal, aparte de especies de *Rafflesia*, para que se haya realizado un esfuerzo especial de conservación en Sabah, y se concluye que el éxito de dicho esfuerzo llevará, a través de un tratamiento similar, a la protección de otros grupos vegetales en peligro.

Part 6

Short communications

The potential for the development of botanic gardens in Madagascar and *ex situ* conservation of Malagasy threatened endemics

Voara Randrianosolo

Department of Natural History
Parc Botanique et Zoologique de Tsimbazaza
Madagascar

Madagascar is the fourth largest island and has been isolated for at least 150 million years. A number of higher plant and animal taxa share this long history of isolation, and as a result offer an insight into the structure of biological communities that have long since disappeared on the contiguous continental African land masses. Macro- and micro-habitat niche diversity derived from the island's spectacular geological and climatic diversity render Madagascar as one of the world's most important living laboratories in evolution.

Unfortunately the natural vegetation has been reduced to less than 20% of its former distribution and is now largely confined to the outer edges of the island. It is essentially divided into three main vegetation divisions, the eastern rain forests, the western deciduous forests and the southern semi-arid spinescent bush. Madagascar's protected area network includes 36 parks and reserves, many of which are experiencing serious degradation as a result of an expanding agricultural frontier, the collection of fuelwood, and uncontrolled burning. The continuation of present trends would undoubtably lead ot the loss of many critical ecosystems within the next few decades.

The Malagasy Government has recently launched the preparation of a national Environmental Action Plan (PAE), as a result of the elaboration of the National Conservation Strategy for sustained development, produced at the International Conference on Conservation of Nature held in Antananarivo in Oct-Nov 1985. One of the principal objectives of the P.A.E. is to assure the conservation of Madagascar's unique

TROPICAL BOTANIC GARDENS: THEIR ROLE IN CONVERVATION AND DEVELOPMENT
ISBN 0—12—346850—7

biological diversity. The Action Plant recommends the creation of plant propagation centers in the six provinces of Madagascar. The rôles played by the only Botanical Garden in Madagascar, a part of Parc de Tsimbazaza, last year, are discussed in term of its close collaboration with university botanical departments in Madagascar and overseas institutions, its taxonomic and field knowledge, its horticultural projects for propagation and maintaining threatened endemic plants in cultivation and its educational program for students and for public in general about why plants are very important in the Malagasy people daily life and how they are in danger. Potential development of satellite botanic gardens of Parc de Tsimbazaza with *ex situ* conservation prospects is being planned, with the opening last year of other universities located in the five provinces other than the capital city, Antananarivo. It is hoped that each one of the satellite gardens could specialize in the conservation of Malagasy threatened endemics of the region where established, as the five provinces correspond to different vegetation types.

The potential for the development of twinning and liaison with botanic gardens in the tropics: a western hemisphere example.

James Affolter*

University of California Botanical Garden
Berkeley, California, U.S.A.

The conservation strategy that emerged from the 1985 Las Palmas Congress stressed the need for further development of existing botanic gardens in the tropics and the potential benefits of twinning between temperate and tropical institutions. In August, 1988, the Robert and Catherine Wilson Botanical Garden (San Vito, Costa Rica) and the University of California Botanical Garden at Berkeley (U.S.A.) agreed to become "Sister Gardens". The decision to formalize our relationship grew out of the success of co-operative efforts during the previous year and a half. The Wilson Garden is a public garden serving a rural agricultural community in southern Costa Rica. It is also a field station for the Organization for Tropical Studies (O.T.S.), a consortium of 40 United States and Costa Rican universities dedicated to education, research, and the wise use of natural resources in the tropics. The University of California Botanical Garden is a public garden as well as a research and educational institution serving the Berkeley campus. It is located in the San Francisco Bay Area, a metropolitan region of more than five million people with a Mediterranean climate. In spite of the superficial differences between the two gardens, a shared focus on research and education and the commitment to plant conservation provide a strong basis for partnership.

A grant from the Stanley Smith Horticultural Trust financed several joint curatorial and educational projects. A computer was purchased for the Wilson Garden along with programmes for computerization of the

---

* Present address: The Cornell Plantations, 100 Judd Falls road, Ithaea, New York 14850, U.S.A.

TROPICAL BOTANIC GARDENS: THEIR ROLE IN CONVERSATION AND DEVELOPMENT
ISBN 0—12—346850—7

plant record system (BG-BASE, distributed and installed by the Center for Plant Conservation). Staff from the U.C. Berkeley Garden made several trips to Costa Rica to discuss curatorial and educational needs with the Wilson Garden director. Many techniques used in the temperate garden transferred easily to the tropical setting. For example, a new map brochure was designed and produced for the Wilson Garden using a recently completed publication from the Berkeley garden as a model. Logistical factors such as the difficulty of communicating over long distances, language differences, and financial constraints pose problems requiring creative solutions but they can be overcome. Future programmes include a course offered through the University of California in July and August, 1989, entitled "Tropical Conservation Education". Students will study curatorial and educational techniques in the U.C. Berkeley Garden before spending six weeks applying their training at the Wilson Botanical Garden and at La Selva Biological Station, another O.T.S. field station in Costa Rica. The "Sister Garden" approach provides one model of how tropical and temperate botanical gardens can develop mutually beneficial programmes by building upon similarities in their objectives. Differences between institutions can be used to mutual advantage, providing the audience and staff of each garden exposure to difference biological, cultural, and socio-economic environments – a window on another world.

Computerized plant record systems for botanic gardens

Kerry S. Walter*

The Center for Plant Conservation
125 The Arborway
Jamaica Plain, MA, U.S.A

One of the primary attributes distinguishing a botanic garden from a purely display garden is the practice of accessioning plants as they enter the collection and then keeping records on this material throughout its life. Without an accurate, up-to-date system of records, a garden's collections can be of little value to its scientific, research, education, or conservation missions. Botanic garden records are inherently complex due to the diversity of information that it is desirable to track for a living collection; chief among these items are scientific and common names, taxonomic placement, immediate, intermediate, and original source data, wild collection data, propagation history, genetic lineages, locations within the collection, special characteristics, and curatorial/maintenance needs and history. Few gardens devote enough time, staff or financial resources to this vital task.

Although the information can be stored and manipulated manually using an accession card and/or accession book, data retrieval is severely limited in such systems, and much information must be entered repetitively. Computerizing these records using relational database management software virtually eliminates entering or storing repetitive data, thus ensuring far greater consistency and accuracy, and vastly reducing the storage requirements. Once computerized, the information can be used not only for curation of the collection, but also for automated production of labels, as a link to computerized mapping systems, and as an educational and interpretative tool itself by means of public-access terminals to the database. Relational database software is now readily

* Present address: World Conservation Monitoring Centre, Cambridge, U.K.

TROPICAL BOTANIC GARDENS: THEIR ROLE IN CONVERVATION AND DEVELOPMENT
ISBN 0—12—346850—7

obtainable for use on relatively inexpensive microcomputers, making this technology available to a much wider number of gardens throughout the world than was the case even a few years ago.

The use of computers per se does not ensure accurate and complete records, but their use can be a tremendous boon for both large and small gardens; most gardens, however, lack the staff or time to design and implement a sophisticated plant records system. Since most gardens need to store the same general kinds of information, having each design its own system with little regard for international compatibility is both wasteful and undesirable. At the request of the Botanic Gardens Conservation Secretariat, a relational database application called BG-BASE has been designed and has now been installed in 25 gardens in four countries.

The Cartagena Botanic Garden, Colombia: an experience in the problems and possibilities of financing a tropical botanic garden

Hermes Cuadros V.

Jardín Botánico Guillermo Piñeres
Cartagena, Colombia

The Republic of Colombia is situated between the coordinates S 41°3'30" N 12°27'46" and 66°50'54", W 79°01'23" in the north of South America. The country is crossed by three mountain ranges of the Andean system, and isolated mountains such as the Sierras of Macarena, Chiribiquete, Baudó, San Lucas, San Jacinto, Cocinas, Jarara and Macuira, and the high elevation of Sierra Nevada of Santa Marta, which rises from sea level up to 5,775 m., and is always covered with snow at the top. Various factors such as the surface of the country (1,141,748 Km^2), its position in the most northern point of South America acting both, as a barrier between the Atlantic and Pacific oceans, and as bridge between North and South America, and the different ecological niches offered by its orography, has enabled the establishment of a great number of plant species in various isolated localities of the Colombian Andes, this being an area of high level of diversity and endemism.

Despite its richness in natural resources, the many necessities of this developing country, have obliged our governments to structure their priorities, and conservation has not yet found a relevant place among these. In third world countries the official bodies do not often have infrastructures that can appropriately channel technical, human, or economic resources, which can help to stop the destruction of nature, due to the extractive activities on which their economy is based.

In developing countries, botanic gardens are the most appropriate institutions to lead the actions related to research and conservation of the environment, if there are not official bodies in charge of this matter. Through my own experience as a Scientific Director of the Botanic

TROPICAL BOTANIC GARDENS: THEIR ROLE
IN CONVERVATION AND DEVELOPMENT
ISBN 0—12—346850—7

Garden "Guillermo Piñeres" and Coordinator of the National Network of Botanic Gardens of Colombia I have observed that the botanic gardens who have been able to carry out their projects most successfully are those established to solve a particular environmental problem and the social necessity of conservation. The botanic gardens in the third world conceived as centres to exhibit ornamental plants have stagnated for many years and languish without providing any effective service to nature conservation. The National Network of Botanic Gardens in Colombia is formed by 15 institutions, divided arbitrarily by myself, in the following four categories:

1 University Botanic Gardens: 5 gardens, attached to the official universities, and under the administrative and financial restrictions of the universities. In Colombia there are no botanic gardens dependent on private universities.

2 Botanic gardens assigned to official bodies: 5 gardens with great potential, autonomy, and enough economical resources and legal status to carry out important conservation tasks. However, they have errors in its conception, and limitations in their managerial staff training. They are also subject to political pressures.

3 Botanic gardens associated to private trusts: there are 3 institutions with autonomy, administrative organization and movility, without bureaucracy, and with basic managerial staff that can develop conservation work, teaching, and their own research, or in association with other national or international organizations.

4 Private botanic gardens: 2 gardens limited by the capital and self-interest of their proprietors.

The Botanic Garden "Guillermo Piñeres" in Cartagena, Colombia, is a modern and very active botanic garden nationally and internationally, with a great potential to develop its conservation programmes, teaching and research, thanks to three favourable factors:

1 It was proposed by a number of influential people of the region as the solution to conservation problems.

2 It has the institutional and financial backup of the Central Bank of Colombia, that is The Bank of The Republic.

3 It has the scientific and unconditional support of the Missouri Botanical Garden.

All this, together with a careful management of its projects and the completion of obligations through the publication of its results, projects an image of security that we hope to spread to all botanic gardens of Latin America.

Resolutions passed at the Second International Botanic Gardens Conservation Congress held on the theme "Tropical Botanic Gardens: Conservation and Development"

ILE DE LA REUNION APRIL 24 - 28, 1989

organized by the:

BOTANIC GARDENS CONSERVATION SECRETARIAT (BGCS) of the INTERNATIONAL UNION FOR CONSERVATION OF NATURE AND NATURAL RESOURCES (IUCN)

LE CONSEIL GENERAL DE LA REUNION - FRANCE

ASSOCIATION FRANCAISE POUR LA CONSERVATION DES ESPECES VEGETALES

LES CONSERVATOIRE ET JARDINS BOTANIQUES DE NANCY - FRANCE

LE CONSERVATOIRE ET JARDIN BOTANIQUE DE MASCARIN - FRANCE

under the patronage of:

LE SECRETARIAT D'ETAT A L'ENVIRONNEMENT (FRANCE)

in technical collaboration with:

THE FOOD AND AGRICULTURE ORGANIZATION OF THE UNITED NATIONS (FAO)

with the participation of:

LA DIRECTION DE LA PROTECTION DE LA NATURE DU SECRETARIAT D'ETAT A L'ENVIRONNEMENT - FRANCE

TROPICAL BOTANIC GARDENS: THEIR ROLE
IN CONVERVATION AND DEVELOPMENT
ISBN 0—12—346850—7

BUREAU DES RESSOURCES GENETIQUES DU MINISTERE DE LA RECHERCHE - FRANCE

L'UNIVERSITE FRANCAISE DE L'OCEAN INDIEN

LA DIRECTION DEPARTEMENTALE DE L'AGRICULTURE ET DES FORETS, REUNION

LA DIRECTION REGIONALE DE L'OFFICE NATIONAL DES FORETS, REUNION

RESOLUTIONS COMMITTEE

Dr E. FORERO (Chairman) - Missouri Botanical Garden, St Louis, Missouri, USA

J.P. GALLAND - Secrétariat d'Etat l'Environnement, Paris

J.C. GIRARD - S.R.E.P.E.N., Saint Denis, Ile de la Réunion

Professor HE SHAN-AN - Nanjing Botanical Garden, Mem. Sun Yat-sen, People's Republic of China

Dra M. CORDEIRO MARINO - Instituto Botanica São Paulo, Brazil

D.B. SUMITHRAARACHCHI - Royal Botanic Gardens, Peradeniya, Sri Lanka

Dr R.L. TAYLOR - Chicago Horticultural Society and Chicago Botanic Garden, Glencoe, Illinois, U.S.A.

B. COE TEIXEIRA - Instituto Botanica São Paulo, Brazil

Dr P.S. WYSE JACKSON - Botanic Gardens Conservation Secretariat (IUCN)

RESOLUTION 1
The Botanic Garden Conservation Strategy

The Second International Botanic Gardens Congress applauds the efforts and contributions made by the IUCN Botanic Garden Conservation Secretariat in the preparation of the valuable and unique document, "The Botanic Gardens Conservation Strategy":

Recognizing the advances made by botanic gardens in the management and establishment of policy guide lines using the recommendations made at the First International Botanic Gardens Congress in Las Palmas, Gran Canaria, recommends that the BGCS make available copies of "The Botanic Gardens Conservation Strategy" to:

1. all the botanic gardens listed in the International Directory compiled by the BGCS;
2. governments, political leaders, international and aid agencies, to facilitate the implementation of plant conservation activities in botanic gardens as a matter of priority and urgency, and,
3. other organizations engaged in germplasm conservation both *in situ* and *ex situ*.

In addition to this, the Congress calls upon the botanic gardens of the world to:

4. act quickly to implement the Botanic Gardens Conservation Strategy in their own gardens;
5. re-examine the mission statements of their gardens in the light of the Strategy recommendations;
6. seek to gain approval for such mission statements throughout all staff and management levels within their institutions;
7. co-operate with the Botanic Garden Conservation Secretariat to help build a strong international network and participate fully in its programme.

RESOLUTION 2
Germplasm Conservation

Noting the considerable and significant recent advances in the conservation of germplasm of cultivated and wild species;

Mindful that priority has hitherto been given by existing germplasm conservation centres to the preservation of the diversity of the world's major crop plants;

Affirming the vital rôle that botanic gardens must play in germplasm conservation;

Acknowledging that over 150 botanic gardens in the world report have seed banks;

Considering that the rapidly expanding technologies and human populations, especially in developing countries, has placed much valuable germplasm at risk;

Recognizing that cultivated varieties/land races are human creations and/or selections and that involvement by indigenous peoples in their inventory and evaluation is essential.

The Congress urges that the Botanic Gardens Conservation Secretariat continue its efforts to:

1. co-ordinate an international network of botanic garden germplasm centres;
2. monitor and promote the development and establishment of new centres for this purpose and foster the growth of existing ones;
3. place particular emphasis on promoting research into and the conservation of useful plants, including germplasm of medicinal species, wild crop relatives, primitive land races of crops, and the so-called minor crops which are not covered by other germplasm networks or organizations;
4. expand the BGCS databases to include details of accessions of such germplasm held in many types of collections, in addition to those in botanic gardens;
5. monitor the movement of germplasm between these centres;

6. undertake efforts to ensure that documentation on rare plant propagation and horticulture is stored and made available as an asset for the botanic garden community.

RESOLUTION 3
Botanic Gardens' Networks, Programmes and Co-operation

Recognizing that botanic gardens in the developing countries of the world should play an effective rôle in plant conservation,

Aware that for these botanic gardens to be effective in implementing conservation strategies they need to plan and carry out programmes that take into account the specific needs of these particular regions,

Realising that botanic gardens need to co-ordinate activities with botanic gardens in other parts of the world and in their own region,

Appreciating that co-ordination is already underway in some countries but not in others; but that networking must be one of the most important functions of these botanic gardens,

The Congress recommends that the Botanic Garden Conservation Secretariat:

1. give special attention towards strengthening existing networks of botanic gardens in developing countries and actively encourage the creation of new ones where they have not yet been formed;
2. encourage the widespread planning, development and implementation of national programmes;
3. encourage regular botanic garden meetings at national and regional levels;
4. encourage general co-operation in sharing resources, training and collection policies;
5. encourage twinning and co-operation between botanic gardens in temperate countries with those in tropical and subtropical countries, including training, general advice, support and technical assistance;
6. undertake to publish a series of regional newsletters; and,

7. provide special emphasis for the aspects mentioned for botanic garden development in the regions of the Western Indian Ocean (including Mascarenes, Madagascar, the Seychelles and Comoros), Latin America, Africa, the Caribbean, Oceania, and the southern and southeast Asia region.

RESOLUTION 4
Botanic Gardens and *in situ* conservation

Considering that botanic gardens are heavily involved in *in situ* conservation of plants and ecosystems and will increasingly become so in the near future;

Agreeing that they need to extend and better plan their activities in this field to cover the many different aspects relating to such conservation;

The Congress recommends that the BGCS actively encourage:

1. research on the restoration of degraded natural environments, including the elimination of invading exotic species, biological control, the rescue, propagation and reintroduction of rare and endangered plants;

2. greater participation of botanic gardens in the management of natural reserves, with greater priority to local species and ecosystems;

3. involvement of botanic gardens in the promotion of the protection of natural environments in their area of influence, including pressure for the establishment of appropriate regulations and legislation;

4. botanic gardens to avoid the introduction of exotic taxa (including hybrids and transgenic plants) to the wild, in order to prevent the spread of new and known biological invaders and pests; and,

5. continuation, with IUCN, of the investigation of the development of data banks on actually or potentially invasive plants.

RESOLUTION 5
Co-operation for Genetic Resource Conservation

Considering that the diversity of plant life represents the indispensable base for meeting the future needs of humanity;

Noting the experience gained by genetic resource centres in the conservation and management of the diversity of major crop plants and the increasing interest by crop genetic resources experts for *in situ* and *ex situ* conservation of wild relatives of crop plants;

The Congress recommends that botanic gardens and their regional and national networks provide similar links with governmental and non-governmental organizations involved in the conservation of genetic resources in their own regions to enable them:

1. to improve training for botanic garden staff;
2. to better co-ordinate and co-operate with other organizations; and,
3. to make their extensive systematic, biogeographic and conservation expertise available to all those promoting the preservation of biological diversity and the environment.

RESOLUTION 6
Plant trade and Botanic Gardens

Recognizing that the present and increasing demand for wild collected plants as a resource for commercial use is of continuing and growing concern for the botanical garden community, as well as the Botanic Gardens Conservation Secretariat and the Secretariat of CITES and to TRAFFIC;

The Congress recommends that the Botanic Gardens Conservation Secretariat:

1. establish and convene a special working group to investigate all aspects of the effective commercialization of plants of conservation concern and present a report to the next Congress; and,
2. initiate co-operative programmes with and between botanic gardens to achieve a more effective distribution of cultivated conservation species, with the aim of providing new financial resources

to support continuing research and conservation on plants in jeopardy in all regions of the world.

RESOLUTION 7
Protection of Primary Tropical Vegetation

Recognizing the heavy loss of primary tropical forests of the world;

Emphasizing that *in situ* conservation of species and ecosystems constitute an essential element of the world conservation strategy and concern of botanic gardens world-wide;

The Congress urges IUCN, all international and national conservation agencies, as well as botanic gardens, to continue to work with governments, local and regional authorities to achieve the immediate and integral protection of viable and sustainable ecosystems of the remaining tropical forests.

RESOLUTION 8
Botanic Gardens and Taxonomy

Recognizing the essential rôle of all aspects of biological science in defining the ways and means of plant conservation;

Noting the present availability of taxonomic expertise amongst the staff of botanic gardens and their associated institutions;

Considering but deploring the present day inadequate funding available for taxonomic and systematic research;

The Congress urges that botanic gardens and their associated institutions place greater emphasis on the future training of taxonomists and support for systematic research.

RESOLUTION 9
3rd International Botanic Gardens Conservation Congress

Considering that the First International Botanic Gardens Conservation Congress held at Las Palmas, Gran Canaria in 1987 recommended that a Botanic Gardens Conservation Congress meet every three years;

Recognizing that both the First and Second Botanic Gardens Conservation Congresses have been held in Old World islands;

The Second International Botanic Gardens Conservation Congress recommends and urges that the BGCS make every effort to hold the Third International Botanic Gardens Conservation Congress in a continental country of high biological diversity in 1992 in the New World.

RESOLUTION 10
Orchids and Conservation

Orchids, the largest family of flowering plants, represent a special position in the world of plants and this Congress recognises that botanic gardens should play a leading rôle in many aspects of their conservation and recommends that:

1. botanists in tropical countries should endeavour to complete check-lists of the Orchidaceae in their countries as soon as possible, including data on distribution and conservation, so that current needs and future policies may be formulated;
2. botanic gardens set an example to amateur, professional and commercial orchid enthusiasts in refraining from collecting specimens from the wild except in rescue operations, or where specific plants are required for ongoing research;
3. the BGCS co-operate with IUCN Species Survival Commission (Orchid Specialists Group) to formulate guidelines for the rescue of living plants;
4. the BGCS encourage botanic gardens to increase their collections of orchids by national and international exchange of pollen, seeds, seedlings, and divisions of plants, together with relevant records of their origins;
5. the BGCS encourage IUCN to continue the evaluation of suggestions for the certification of orchid nurseries which have already been proposed by the International Orchid Commission and, if successful, this scheme should also be extended to other groups of threatened plants of horticultural significance.

RESOLUTION 11
Public Education in Botanic Gardens

Considering the responsibilities and experience gained in botanic gardens in the field of research, education, and display of plants and their environmental needs, the Congress urges all botanic gardens to expand their public education programmes for promoting plant conservation knowledge and ethics for visitors, especially for school children, local communities, and regional authorities.

Furthermore, recognizing that botanic gardens must continue to place greater emphasis on the scientific management and monitoring of the plants in their collections, the Congress recommends that all botanic gardens should endeavour to communicate the valuable research aspects of their conservation work to the widest possible audience, and thus enhance the reputation of botanic gardens for conservation.

RESOLUTION 12

The participants in the 2nd International Botanic Gardens Conservation Congress are profoundly moved by the warm welcome and hospitality they have received in the Island of Réunion. They wish to express their most sincere thanks to all those involved in the organization of this important Congress, particularly to the Conseil Général de la Réunion for its generous support, both financially and in other ways; to the French Minister of the Environment, for the interest that he has shown in this meeting, and to the staff of the Conservatoire et Jardin Botanique de Mascarin for their remarkable commitment and work in making this Congress such a success.

RESOLUTIONS adoptées par LE DEUXIEME CONGRES INTERNATIONAL DES JARDINS BOTANIQUES POUR LA CONSERVATION sur le thème "Jardins Botaniques Tropicaux: Conservation et Développement"

ILE DE LA REUNION 24 - 28 AVRIL, 1989

organisé par:

SECRETARIAT DES JARDINS BOTANIQUES POUR LA CONSERVATION (BGCS) de la INTERNATIONAL UNION FOR CONSERVATION OF NATURE AND NATURAL RESOURCES (IUCN)

LE CONSEIL GENERAL DE LA REUNION - FRANCE

ASSOCIATION FRANCAISE POUR LA CONSERVATION DES ESPECES VEGETALES

LES CONSERVATOIRE ET JARDINS BOTANIQUES DE NANCY - FRANCE

LE CONSERVATOIRE ET JARDIN BOTANIQUE DE MASCARIN - FRANCE

sous le haut patronage de:

LE SECRETARIAT D'ETAT A L'ENVIRONNEMENT (FRANCE)

en collaboration technique avec:

THE FOOD AND AGRICULTURE ORGANIZATION OF THE UNITED NATIONS (FAO)

TROPICAL BOTANIC GARDENS: THEIR ROLE IN CONVERVATION AND DEVELOPMENT
ISBN 0—12—346850—7

et avec la participation de:

LA DIRECTION DE LA PROTECTION DE LA NATURE DU SECRETARIAT D'ETAT A L'ENVIRONNEMENT - FRANCE

BUREAU DES RESSOURCES GENETIQUES DU MINISTERE DE LA RECHERCHE - FRANCE

L'UNIVERSITE FRANCAISE DE L'OCEAN INDIEN

LA DIRECTION DEPARTEMENTALE DE L'AGRICULTURE ET DES FORETS, REUNION

LA DIRECTION REGIONALE DE L'OFFICE NATIONAL DES FORETS, REUNION

RESOLUTION 1
La stratégie de conservation pour les jardins botaniques

Le Second Congrès International des Jardins Botaniques applaudit aux efforts et aux contributions du Secrétariat des Jardins Botaniques pour la Conservation (BGCS) pour la réalisation du document unique et précieux intitulé "La Stratégie des Jardins Botaniques pour la Conservation".

Reconnaissant les progrès fait par les jardins botaniques dans l'élaboration et la mise en oeuvre des grandes lignes d'une politique basée sur les recommandations du Premier Congrès International des Jardins Botaniques à Las Palmas (Grande Canarie), il recommande que le BGCS diffuse des exemplaires de "La Stratégie des Jardins Botaniques pour la Conservation":

1. à tous les jardins botaniques qui apparaissent dans le Répertoire International réalisé par le BGCS;
2. aux gouvernements, responsables politiques, organismes internationaux et agences d'aide, pour faciliter la mise en oeuvre d'actions de conservation des plantes dans les jardins botaniques de façon urgente et prioritaire, et
3. aux autres organisations impliquées dans la conservation des ressources génétiques tant *in situ* que *ex situ*.

En outre, le Congrès invite les jardins botaniques du monde à:

4. agir au plus vite pour mettre en oeuvre la stratégie de conservation pour les jardins botaniques dans leurs propres jardins;
5. réexaminer les objectifs assignés à leurs jardins à la lumière des recommandations de la stratégie;
6. rechercher l'approbation sur ces objectifs de la part de tout le personnel et de la direction de leurs institutions;
7. coopérer avec le Secrétariat des Jardins Botaniques pour la Conservation pour mettre sur pied un réseau international puissant, et participer pleinement à son programme.

RESOLUTION 2
Conservation des ressources génétiques

Constatant les progrès récents et considérables réalisés dans la conservation des ressources génétiques des espèces cultivées et sauvages;

Soucieux du fait que la priorité ait été donnée jusqu'à maintenant par les centres de ressources génétiques existants à la préservation de la diversité des plantes agricoles d'importance mondiale;

Affirmant le rôle vital que les jardins botaniques doivent jouer dans la conservation des ressources génétiques;

Reconnaissant le fait que plus de 150 jardins botaniques dans le monde déclarent avoir des banques de graines;

Considérant que la croissance rapide des technologies et des populations humaines, surtout dans les pays en voie de développement, a mis en danger beucoup de ressources génétiques précieuses;

Reconnaissant que les variétés cultivées sont des créations et/ou des sélections humaines et que l'implication des peuples indigènes dans leur inventaire et leur évaluation est essentielle, le Congrès presse le Secrétariat des Jardins Botaniques pour la Conservation à continuer ses efforts pour:

1. coordonner un réseau international de centres de ressources génétiques dans les jardins botaniques;

2. superviser et promouvoir le développement et l'établissement de centres nouveaux ayant cet objectif, et stimuler la croissance des centres existants;
3. porter une attention particulière à la promotion de la recherche et de la conservation des plantes utiles, y compris les plantes médicinales, les variétés locales primitives, et les plantes vivrières dites mineures qui ne sont pas couvertes par d'autres réseaux ou organisations de ressources génétiques;
4. élargir les bases de données du BGCS de façon à inclure les données sur les accessions de ce type de matériel détenues dans les diverses collections, en plus de celles des jardins botaniques;
5. superviser les échanges de matériel génétique entre ces centres;
6. faire en sorte que la documentation sur la propagation et la culture des plantes rares soit rassemblée et rendue disponible à la communauté des jardins botaniques.

RESOLUTION 3
Réseaux, programmes et coopération entre jardins botaniques

Reconnaissant que les jardins botaniques dans les pays du monde en voie de développement devraient jouer un rôle effectif dans la conservation des plantes;

Conscient que pour mettre réellement en oeuvre des stratégies de conservation, ces jardins botaniques doivent préparer et réaliser des programmes qui prennent en compte les besoins spécifiques de leurs régions;

Reconnaissant que les jardins botaniques ont besoin de se coordonner avec ceux des autres parties du monde et de leur propre région;

Appréciant que la coordination soit déjà réalisée dans certains pays mais pas dans d'autres; mais que la constitution de réseaux doit être une des fonctions les plus importantes de ces jardins botaniques;

Le Congrès recommande que le Secrétariat des Jardins Botaniques pour la Conservation:

1. accorde une attention particulière au renforcement des réseaux existants de jardins botaniques dans les pays en voie de développement, et encourage activement la création de nouveaux réseaux là ou ils n'existent pas encore;
2. encourage de la façon la plus large l'élaboration et la mise en oeuvre de programmes nationaux;
3. encourage des rencontres régulières entre jardins botaniques aux niveaux national et régional;
4. encourage une coopération générale dans le partage des ressources et les politiques de formation et de collections;
5. encourage le jumelage et la coopération entre les jardins botaniques des pays tempérés et ceux des pays tropicaux et sub-tropicaux, dans le domaine de la formation, du conseil, du soutien et de l'assistance technique;
6. entreprenne la publication d'une série de lettres d'information régionales; et
7. porte une attention particulière à ces aspects pour le développement des jardins botaniques dans l'Ouest de l'Océan Indien (Mascareignes, Madagascar, Seychelles et Comores), l'Amérique Latine, les Antilles, l'Océanie et l'Asie du Sud et du Sud-Ouest.

RESOLUTION 4
Les jardins botaniques et la conservation *in situ*

Considérant que les jardins botaniques sont fortement engagés dans la conservation *in situ* des plantes et des écosystèmes et le deviendront de plus en plus dans un futur proche;

Reconnaissant qu'il leur faut étendre et mieux planifier leurs actions dans ce domaine de façon à couvrir les aspects très divers de ce type de conservation;

Le Congrès recommande que le BGCS encourage activement:

1. la recherche sur la restauration des environnements naturels dégradés, ce qui inclut l'élimination des espèces exotiques envahis-

santes, le contrôle biologique, le sauvetage, la propagation et la réintroduction des plantes rares et menacées;

2. une plus grande participation des jardins botaniques dans la gestion des réserves naturelles, avec une priorité accrue pour les espèces et écosystèmes locaux;
3. l'engagement des jardins botaniques dans la promotion de la protection des environnements naturels de leur zone d'influence, et en particulier la pression pour que soient édictés les lois et règlements appropriés;
4. les jardins botaniques à éviter l'introduction de taxons exotiques (y compris les hybrides et les plantes transgéniques) dans la nature, de façon à prévenir la diffusion d'envahisseurs et de pestes biologiques connus ou nouveaux; et
5. la poursuite, avec l'UICN, de la recherche et du développement de bases de données sur les plantes réellement ou potentiellement envahisseuses.

RESOLUTION 5
Coopération pour la conservation des ressources génétiques

Considérant que la diversité végétale représente la base indispensable pour subvenir aux besoins futurs de l'humanité;

Notant l'expérience acquise par les centres de ressources génétiques dans la conservation et la gestion de la diverité des principales plantes cultivées et l'intérêt croissant des spécialistes des ressources génétiques agricoles pour la conservation *in situ* et *ex situ* des plantes sauvages apparentées aux plantes cultivées;

Le Congrès recommande que les jardins botaniques et leurs réseaux régionaux et nationaux se rapprochent des organisations gouvernementales ou non-gouvernementales engagées dans la conservation des ressources génétiques dans leur région, afin de leur permettre:

1. d'améliorer la formation du personnel des jardins botaniques;
2. de mieux coordonner leurs actions avec les autres organisations;

3. de rendre disponible leur expertise étendue en matière de systématique, de biogéographie et de conservation à tous ceux qui agissent pour la préservation de la diversité biologique et de l'environnement.

RESOLUTION 6
Commerce de plantes et jardins botaniques

Reconnaissant que la demande actuelle croissante de plantes récoltées dans la nature en tant que ressource pour un usage commercial est une préoccupation permanente et grandissante pour la communauté des jardins botaniques, ainsi que pour le Secrétariat des Jardins Botaniques pour la Conservation, le secrétariat de la CITES et le TRAFFIC;

Le Congrès recommande que le Secrétariat des Jardins Botaniques pour la Conservation:

1. mette en place et réunisse un groupe de travail spécial pour étudier tous les aspects de la commercialisation effective des plantes concernées par la conservation, et présente un rapport au prochain Congrès; et
2. lance des programmes avec les jardins botaniques pour arriver à une distribution plus efficace des espèces menacées qui sont en culture, afin de fournir des ressources financières nouvelles pour soutenir la recherche et la conservation des plantes menacées dans toutes les régions du monde.

RESOLUTION 7
Protection de la végétation tropicale primaire

Reconnaissant la forte régression des forts tropicales primaires dans le monde entier;

Soulignant que la conservation *in situ* des espèces et des écosystèmes constitue un élément essentiel de la stratégie mondiale de conservation et une préoccupation des jardins botaniques du monde entier;

Le Congrès presse l'UICN, toutes les agences internationales et nationales de conservation, ainsi que les jardins botaniques, à continuer à agir avec les Etats et les autorités locales et régionales pour aboutir à la protection immédiate et intégrale d'écosystèmes viables et durables dans les forêts tropicales restantes.

RESOLUTION 8
Les jardins botaniques et la taxinomie

Reconnaissant le rôle essentiel de tous les aspects de la science biologique pour définir les voies et moyens de la conservation des plantes;

Notant la disponibilité actuelle de l'expertise taxinomique parmi le personnel des jardins botaniques et des institutions qui leur sont liées;

Considérant tout en la déplorant l'insuffisance des financements disponibles pour la recherche taxinomique et systématique;

Le Congrès presse les jardins botaniques et les institutions qui leur sont liées à accorder davantage d'attention à la formation de taxinomistes et au soutien à la recherche en systématique.

RESOLUTION 9

Troisième Congrès International des Jardins Botaniques pour la Conservation

Considérant que le Premier Congrès International des Jardins Botaniques pour la Conservation tenu à Las Palmas (Grande Canarie) en 1985 a recommandé qu'un Congrès des Jardins Botaniques pour la Conservation soit organisé tous les trois ans;

Reconnaissant que le Premier comme le Second Congrès des Jardins Botaniques pour la Conservation se sont tenus dans des îles de l'Ancien Monde;

Le Second Congrès International des Jardins Botaniques pour la Conservation recommande et presse le BGCS à tout faire pour que le Troisième Congrès International des Jardins Botaniques pour la Conservation se réunisse dans un pays continental de grande diversité biologique en 1992 dans le Nouveau Monde.

RESOLUTION 10
Orchidées et conservation

Les Orchidées, qui sont la famille de plantes à fleurs la plus importante, ont une position spéciale dans le monde des plantes, et ce Congrès reconnaît que les jardins botaniques devraient jouer un rôle de premier plan dans de nombreux aspects de leur conservation, et recommande que:

1. les botanistes des pays tropicaux s'efforcent de compléter les listes d'Orchidaceae de leurs pays au plus vite, en incluant les données sur leur distribution et leur conservation, de façon à ce que les besoins actuels et les politiques futures puissent être formulés;
2. les jardins botaniques montrent l'exemple aux passionnés d'orchidées, amateurs, professionnels ou commerçants, en s'abstenant de récolter des spécimens dans la nature, sauf lors d'opérations de sauvetage, ou quand des plantes spécifiques sont nécessaires pour des recherches en cours;
3. le BGCS coopère avec la Commission pour la Survie des Espèces de l'UICN (Groupe des Spécialistes d'Orchidées) pour élaborer des normes pour le sauvetage de plantes vivantes;
4. le BGCS encourage les jardins botaniques à accroître leurs collections d'orchidées par des échanges nationaux et internationaux de pollen, semences, plantes et éclats de plantes, accompagnés des données correspondantes sur les origines;
5. le BGCS encourage l'UICN à continuer l'évaluation des suggestions pour la certification de pépinières d'orchidées qui ont déjà été proposées par la Commission Internationale des Orchidées et, en cas de succès, à étendre ce schéma à d'autres groupes de plantes menacées d'importance horticole.

RESOLUTION 11
Education du public dans les jardins botaniques

Considérant les responsabilités et l'expérience acquises par les jardins botaniques dans les domaines de la recherche, de l'éducation et de la présentation des plantes et de leurs besoins écologiques, le Congrès presse tous les jardins botaniques à élargir leurs programmes d'éducation du public pour promouvoir la connaissance et l'éthique sur la conservation des plantes chez leurs visiteurs, en particulier les enfants d'âge scolaire, les communautés locales et les autorités régionales.

De plus, reconnaissant que les jardins botaniques doivent continuer à accorder plus d'attention à la gestion et au contrôle scientifique de leurs collections végétales, le Congrès recommande que tous les jardins botaniques s'efforcent de communiquer les aspects de recherche importants de leur action de conservation au public le plus large, et par là-même d'améliorer la réputation des jardins botaniques dans la conservation.

RESOLUTION 12

Les participants du Second Congrès International des Jardins Botaniques pour la Conservation ont été profondément touchés par l'accueil chaleureux et l'hospitalité qu'ils on reçus à l'Ile de la Réunion. Ils souhaitent exprimer leurs plus sincères remerciements à toutes les personnes impliquées dans l'organisation de cet important Congrès, et spécialement au Conseil Général de la Réunion pour son soutien généreux sur tous les plans; au Ministre Françaisde l'Environnement pour l'intérêt qu'il a montré à cette réunion, ainsi qu'au personnel du Conservatoire et Jardin Botanique du Mascarin pour leur engagement et leur travail remarquables qui ont assuré le succès de ce Congrès.

RESOLUCIONES promulgadas en EL SEGUNDO CONGRESO INTERNACIONAL DE JARDINES BOTANICOS Y LA CONSERVACION que trató el tema "Jardines Botánicos Tropicales: Conservación y Desarrollo"

ISLA DE LA REUNION ABRIL 24-28, 1989

organizado por

LA SECRETARIA PARA LA CONSERVACION EN JARDINES BOTANICOS (BGCS) de la INTERNATIONAL UNION FOR CONSERVATION OF NATURE AND NATURAL RESOURCES (IUCN)

LE CONSEIL GENERAL DE LA REUNION - FRANCE

ASSOCIATION FRANCAISE POUR LA CONSERVATION DES ESPECES VEGETALES

LES CONSERVATOIRE ET JARDINS BOTANIQUES DE NANCY - FRANCE

LE CONSERVATOIRE ET JARDIN BOTANIQUE DE MASCARIN - FRANCE

bajo el patrocinio del

SECRETARIAT D'ETAT A L'ENVIRONNEMENT (FRANCE)

con la colaboración técnica de

THE FOOD AND AGRICULTURE ORGANIZATION OF THE UNITED NATIONS (FAO)

TROPICAL BOTANIC GARDENS: THEIR ROLE IN CONVERVATION AND DEVELOPMENT
ISBN 0—12—346850—7

con la participación de

LA DIRECTION DE LA PROTECTION DE LA NATURE DU SECRETARIAT D'ETAT A L'ENVIRONNEMENT - FRANCE

BUREAU DES RESSOURCES GENETIQUES DU MINISTERE DE LA RECHERCHE - FRANCE

L'UNIVERSITE FRANCAISE DE L'OCEAN INDIEN

LA DIRECTION DEPARTEMENTALE DE L'AGRICULTURE ET DES FORETS, REUNION

LA DIRECTION REGIONALE DE L'OFFICE NATIONAL DES FORETS, REUNION

RESOLUCION 1
La Estrategia de Jardines Botánicos para la Conservación

El Segundo Congreso Internacional de Jardines Botánicos aplaude los esfuerzos y las contribuciones que ha hecho la Secretaría para la Conservación en Jardines Botánicos para preparar el valioso y único documento titulado "La Estrategia de Jardines Botánicos para la Conservación":

Reconociendo los avances hechos por los jardines botánicos en el manejo y establecimiento de políticas basadas en las recomendaciones del Primer Congreso Internacional de Jardines Botánicos de Las Palmas, Gran Canaria,recomienda que la BGCS haga llegar copias de "La Estrategia de Jardines Botánicos para la Conservación" a:

1. todos los jardines botánicos incluídos en el Directorio Internacional preparado por la BGCS;

2. gobiernos, líderes políticos, entidades internacionales y financiadoras para facilitar la implementación de actividades de conservación de plantas en jardines botánicos como un asunto prioritario y urgente; y

3. otras organizaciones involucradas en la conservación de germoplasma tanto *in situ* como *ex situ*.

Además, el Congreso solicita a los jardines botánicos que:

4. actúen rápidamente para implementar la Estrategia de Jardines Botánicos para la Conservación;
5. reexaminen sus declaraciones de misión con base en las recomendaciones de la Estrategia;
6. intenten obtener la aprobación de tales declaraciones de misión a todos los niveles administrativos dentro de sus instituciones;
7. cooperen con la Secretaría para la Conservación en Jardines Botánicos para ayudar a construir una red potente internacional y que participen activamente en su programa.

RESOLUCION 2
Conservación de germoplasma

Notando los considerables e importantes adelantos recientes en la conservación de germoplasma de especies cultivadas y silvestres;

Teniendo en cuenta la prioridad dada hasta ahora en centros de conservación de germoplasma ya existentes a la preservación de la diversidad de las plantas cultivadas más importantes del mundo;

Afirmando el papel vital que tienen que jugar los jardines botánicos en la conservación de germoplasma;

Reconociendo que más de 150 jardines botánicos incluídos en el informe mundial tienen bancos de semillas;

Considerando que las tecnologías modernas y el rápido crecimiento de las poblaciones humanas, especialmente de los países en desarrollo, han puesto en peligro germoplasma muy valioso;

Reconociendo que las variedades cultivadas son creaciones y/o selecciones del hombre y que es esencial que las poblaciones indígenas participen en su inventorio y evaluación;

El Congreso urge a la Secretaría para la Conservación en Jardines Botánicos para que continúe sus esfuerzos con el fin de:

1. coordinar una red internacional de centros de germoplasma en los jardines botánicos;
2. dirigir y promover el desarrollo y establecimiento de nuevos centros dedicados a ese fin y estimular el crecimiento de los existentes;

3. enfatizar particularmente en promover la investigación y conservación en plantas útiles, entre las cuales se incluyen el germoplasma de especies medicinales, parientes silvestres, variedades primitivas de cultivos silvestres y los cultivos llamados menores que estén incluídos en otras redes u organizaciones de germoplasma;
4. ampliar las bases de datos de la BGCS para incluir detalles de adquisición de germoplasma contenido en muchos tipos de colecciones, además de las de los jardines botánicos;
5. dirigir el movimiento de germoplasma entre esos centros;
6. hacer esfuerzos para asegurar que la documentación relacionada con la propagación y la horticultura de plantas raras esté almacenada y disponible como un recurso importante para la comunidad de jardines botánicos.

RESOLUCION 3
Redes, programas y cooperación entre jardines botánicos

Reconociendo que los jardines botánicos en los países en desarrollo deben jugar un papel efectivo en la conservacíon de plantas;

Comprendiendo que para implementar estrategias efectivas de conservación,estos jardines botánicos deben planificar y adelantar programas que tomen en cuenta las necesidades específicas de estas regiones particulares;

Reconociendo que los jardines botánicos necesitan coordinar actividades con jardines botánicos en otras partes del mundo y en su propia región;

Comprendiendo que ya existe coordinación en algunos países pero no en otros,pero que el proceso de organizar redes debe ser una de las metas más imporantes de estos jardines botánicos;

El Congreso recomienda que la Secretaría para la Conservación en Jardines Botánicos:

1. preste atención especial al fortalecimiento de redes de jardines botánicos ya existentes en países en desarrollo y estimule activamente la creación de otras redes donde aún no se hayan formado.

2. estimule ampliamente la planificación, el desarrollo y la implementación de programas nacionales;
3. estimule reuniones regulares de jardines botánicos a nivel nacional y regional;
4. estimule la cooperación general para compartir recursos, entrenamiento y políticas de colección;
5. estimule el establecimiento de jardines "gemelos" y la cooperación entre los jardines botánicos de países templados y los de países tropicales y subtropicales, lo cual debe incluir entrenamiento, asesoría general, apoyo y ayuda técnica;
6. se empene en publicar una serie de boletines regionales; y
7. enfatizar especialmente sobre aspectos arriba mencionados , en el desarrollo de jardines botánicos en el Océano Indico Occidental (que incluye las Mascarenes, Madagascar, las Seychelles y los Comoros), América Latina, Africa, el Caribe, Oceania y regiones del sur y sureste de Asia.

RESOLUCION 4
Los jardines botánicos y la conservación *in situ*

Considerando que los jardines botánicos están muy comprometidos en la conservación *in situ* de plantas y ecosistemas y que lo estarán aún más en un futuro cercano;

acordando en que necesitan extender y planear mejor sus actividades en este campo para incluir los diversos aspectos relacionados con tal conservación;

El Congreso recomienda que la BGCS estimule activamente:

1. investigaciones sobre la restauración de ambientes naturales degradados incluyendo la eliminación de especies exóticas invasoras, el control biológico, el rescate, la propagación y la reintroducción de plantas raras y amenazadas:
2. más participación de los jardines botánicos en el manejo de reservas naturales, dando prioridad a especies y ecosistemas locales;

3. la participación de los jardines botánicos en la promoción de la protección de ambientes naturales en su área de influencia, lo cual incluye ejercer presión para lograr el establecimiento de reglamentos y legislación apropiados;

4. a los jardines botánicos para que eviten la introducción de taxa exóticos (se incluyen plantas híbridas y transgénicas) en ambientes silvestres, con el fin de prevenir la difusión de plantas invasoras y plagas biológicas nuevas y conocidas;

5. la continuación, con IUCN, de la exploración del desarrollo de bancos de datos sobre plantas real o potencialmente invasoras.

RESOLUCION 5
Cooperación para la conservación de recursos genéticos

Considerando que la diversidad de recursos vegetales representa la base imprescindible para responder a las necesidades futuras de la raza humana;

Notando la experiencia ganada por los centros de recursos genéticos en la conservación y el manejo de la diversidad de cultivos mayores y el interés creciente de los expertos en recursos genéticos sobre la conservación *in situ* y *ex situ* de parientes silvestres de plantas cultivadas;

El Congreso recomienda que los jardines botánicos y sus redes regionales y nacionales provean vínculos semejantes con organizaciones gubernamentales y no gubernamentales que participan en la conservación de los recursos genéticos de sus propias regiones para ayudarlos a:

1. mejorar el entrenamiento del personal de los jardines botánicos;

2. coordinar y cooperar mejor con otras organizaciones; y

3. hacer disponible su extenso conocimiento sistemático, biogeográfico y conservacionista a todos los que están promoviendo la preservación de la diversidad biológica y el medio ambiente.

RESOLUCION 6
La comercialización de plantas y los jardines botánicos

Reconociendo que la demanda actual y cada vez mayor de colecciones de plantas silvestres como un recurso para uso comercial es una preocupación contínua tanto de la comunidad de jardines botánicos como de la Secretaría para la Conservación en Jardines Botánicos, la Secretaría de CITES y TRAFFIC;

El Congreso recomienda que la Secretaría para la Conservación en Jardines Botánicos:

1. establezca y convoque a un grupo especial de trabajo para investigar todos los aspectos de efectiva comercialización de plantas de importancia para la conservación y para presentar un informe al próximo Congreso; e

2. inicie programas cooperativos con los jardines botánicos para alcanzar una distribución más eficaz de especies cultivadas, con el fin de proveer nuevos recursos financieros para apoyar la investigación contínua sobre plantas en peligro y su conservación permanente en todas partes del mundo.

RESOLUCION 7
La protección de la vegetación primaria tropical

Reconociendo la inmensa pérdida de bosques tropicales primarios en todo el mundo;

enfatizando en el hecho de que la conservación *in situ* de especies y ecosistemas constituye un elemento esencial de la estrategia mundial para la conservación y preocupación de los jardines botánicos en todo el mundo;

El Congreso urge tanto a la IUCN y a todas las organizaciones conservacionistas internacionales y nacionales como también a los jardines botánicos para que continúen trabajando con gobiernos, entidades locales y regionales para alcanzar la protección inmediata e íntegra de ecosistemas viables y sostenibles en los bosques tropicales restantes.

RESOLUCION 8
Los jardines botánicos y la taxonomía

Reconociendo el papel esencial de todos los aspectos de las ciencias biológicas en la definición de los métodos y sistemas de la conservación de plantas;

Notando la disponibilidad actual de conocimento taxonómico dentro del personal de los jardines botánicos y de sus instituciones asociadas;

Considerando pero deplorando los fondos inadecuados disponibles en la actualidad para la investigación taxonómica y sistemática;

El Congreso urge a los jardines botánicos y a sus instituciones asociadas a que den mayor énfasis al entrenamiento de taxónomos y al apoyo de investigaciones sistemáticas.

RESOLUCION 9
Tercer Congreso Internacional de Jardines Botánicos y la Conservación

Considerando que el Primer Congreso Internacional de Jardines Botánicos y la Conservación realizado en Las Palmas en 1985 recomendó que se convocara a un Congreso de Jardines Botánicos y la Conservación cada tres años; Reconociendo que tanto el Primero como el Segundo Congreso de Jardines Botánicos y la Conservación se efectuaron en islas del Viejo Mundo:

El Segundo Congreso Internacional de Jardines Botánicos y la Conservación recomienda y urge que la BGCS se esfuerze en organizar para 1992 el Tercer Congreso Internacional de Jardines Botánicos y la Conservación en un país continental de alta diversidad biológica en el Nuevo Mundo.

RESOLUCION 10
Las orquídeas y la conservación

Las orquídeas, la familia más grande de plantas con flores, tienen una posición especial en el mundo vegetal y este Congreso reconoce que los jardines botánicos deberían jugar un papel de liderazgo en muchos aspectos de su conservación y recomienda que:

1. los botánicos en países tropicales deberían dedicarse a completar listas anotadas de las orquideas sus propios países a la brevedad posible; éstas deben incluir datos sobre la distribución y conservación para que las necesidades actuales y las políticas futuras puedan ser formuladas;
2. los jardines botánicos sirvan de ejemplo para que los entusiastas de las orquídeas, sean aficionados, profesionales o comerciales,se abstengan en coleccionar especímenes silvestres excepto en operaciones de rescate o cuando se necesiten plantas específicas para investigaciones en progreso;
3. la BGCS coopere con la Comisión para la Supervivencia de las Especies (Grupo de Especialistas en Orquídeas) para formular guías para el rescate de plantas vivas;
4. la BGCS estimule a los jardines botánicos a que aumenten sus colecciones de orquídeas mediante el canje internacional de pólen, semillas, plántulas y partes de plantas junto con anotaciones referentes a sus orígenes;
5. la BGCS insista ante la IUCN para que continúe la evaluación de sugerencias para la certificación de criaderos de orquídeas que ya han sido propuestos por la Comisión Internacional de Orquídeas y, de obtener éxito, este plan debería extenderse a otros grupos de plantas amenazadas que sean importantes para la horticultura.

RESOLUCION 11
La educación pública en los jardines botánicos

Considerando las responsabilidades y la experiencia ganadas en los jardines botánicos en los campos de investigación, educación y exhibición de plantas y sus necesidades ambientales, el Congreso inste a todos los jardines botánicos para que amplien sus programas de educación pblica para promover elconocimiento y la ética de la conservación de plantas a los visitantes,especialmente para estudiantes, comunidades locales y autoridades regionales.

Además, reconociendo que los jardines botánicos deben continuar haciendo másénfasis en el manejo científico y en el monitoreo de las plantas en sus colecciones, el Congreso recomienda que todos los jardines botánicos se empenen en difundir a la más amplia audiencia posible los aspectos valiosos investigativos de sus trabajos conservacionistas, con el fin de realizar la reputación de los jardines botánicos en conservación.

RESOLUCION 12

Los participantes en el Segundo Congreso Internacional de Jardines Botánicos y la Conservación están profundamente agradecidos por la amable acogida y por la hospitalidad que recibieron en la Isla de Réunion. Así mismo, desean expresar su más sincero reconocimiento a quienes participaron en la organización de este importante Congreso, y especialmente al Consejo General de la Réunion por su generoso apoyo tanto financiero como de otros tipos; al Ministro Francés del Medio Ambiente por el interés que ha demostrado en esta reunión; y al personal del Conservatoire et Jardin Botanique de Mascarin, por su extraordinaria dedicación y por el trabajo que realizaron para que este Congreso tuviese tanto exito.

Index

A. Camus *see Cupressus dupreziana*
Acacia aff. *pennata* 200
Acampe rigida 200, 203
Achyrospermum seychellarum 198
Aldabra
  endemic species 207
  indigenous species 199
  native species 207
  Seychelles and threatened flora 206
alien plants, invasive 178
alien plants, *see also* indigenous/alien plants
Amaracarpus pubescens 200
Andean flora 89–94
Angiopteris angustifolia 231
Angraecum eburneum 203
Angraecum maheense 198
Atlantic forest conservation and Rio de Janeiro botanical garden rôle 163–6
Australia and botanic gardens 35–41

Bakerella clavata 195, 203
BGCS *see* Botanic Gardens Conservation Secretariat
Bogor Botanic Gardens and orchid conservation 297–305
Botanic Gardens Conservation Secretariat 49–50
Brazil
  Atlantic forest 163–6
  university botanic garden development 119–24
  collection/cultivation of rare native tree species 122–3
  land use 121–2
  orientation programme 122
Brexia madagascariensis 201

Calarcá project 92–4
Carissa edulis 197–8, 202
Cartagena Botanic Garden (Colombia) 337–9
case histories in Mauritius/Rodrigues and *ex situ* collections 100–3
case studies of tropical botanical gardens 79–166
Cassipourea lanceolata 201
CITES and plant trade 249–64
Coffea spp. 102
collection/cultivation of rare native tree species in Brazil 122–3
Colombia *see* Cartagena Botanic Garden community development and ethnobotanical garden/tropical plant conservation 83–4
conservation
  activities, *ex situ* 64–8
  ethnobotanical garden and tropical plant conservation 83
  long-term and orchids 311–14
  measures and fernarium 226–7
  measures in Galápagos islands 184–6
  short-term and orchids 314–16
constraints in Sri Lanka and medicinal plants conservation 133
Côte d'Azur and tropical/sub-tropical trees, rare or threatened 151–60
Crinum mauritianum 101
Crotalaria laburnoides 201
cultivation 65–6
  important plants in *ex situ* collections 103–4
  propagation in fernarium 230–2
Cupressus dupreziana 156–60
  botanical characteristics 156–8
  natural environment 158
  protection and collection 158–60

Dendrobium secundum (orchid fruit) 284
Dioscorea bemarivensis 204
distribution of introduced mammals in Galápagos islands 177

education in ethnobotanical garden and tropical plant conservation 81–2
education and fernarium 228
Ehretia corymbosa 202
endangered plants/species
  Lipizauga botanical sanctuary 245
  Mauritius 107–12
  orchid propagation 292–4
  Rodrigues 113–14
endemic species
  Aldabra 207
  granite islands 195–9, 207
  Malagasy 331–2

Eria pubescens 280
ethnobotanical garden and tropical plant conservation 79–85
community development 83–4
education 81–2
research 80–1
Eugenia capensis 202
Eulophidium pulchrum 200
Eulophidium seychellarum 195
ex situ collections in Mauritius and Rodrigues 95–105
ex situ conservation
France and wild flora protection 64–8
Galápagos islands 88–9
Malagasy 331–2
Seychelles and threatened flora 206
extinct (apparently) species in granite islands 207
extinct species in Seychelles and threatened flora 194–5

fern flora, original in fernarium 229–30
fern-allies 232–7
threatened 225–6
fernarium, rôle of in Malaysia 223–38
aesthetic value 228
concept 227
conservation measures 226–7
cultivation and propagation 230–2
educational programme 228
fern flora, original 229–30
fern-allies, threatened 225–6
ferns and fern-allies grown 232–7
ferns threatened 225–6
project 228–9
propagation techniques 237–8
scientific programme 227–8
vanishing forests 224–6
ferns
Fernarium (Malaysia) 232–7
ornamental 232
rare/threatened 223–38
threatened 225–6
Ficus laterifolia 102
finance and staff training in tropics 50–2
finance of tropical botanic gardens 337–9
flora, indigenous 171–3
foreign genes prevention of 65
forests, vanishing 224–6
Fort Napoléon and Jardin Exotique 72–4
France *see* TRAFFIC-France
France and wild flora protection 59–75
ex situ conservation activities 64–8
figures 59–60
identification of threatened species 60–1
species protection, legal measures for 61–2
wild species protection, legal measures for 62–3
France and WWF *see* WWF-France
French Mediterranean coast *see* Côte d'Azur

Galápagos islands
distribution of introduced mammals 177
ex situ conservation element 88–9
indigenous flora 171–3
indigenous hardwood trees 182–3
indigenous vegetation 173–4
indigenous/alien plants 169–89
colonized zones 183
conservation measures, current 184–6
future prospects 186–8
land use practices and invasive organisms 180–1
local economy and development rôle 181–3
threats 175–80
watersheds 183–4
invasive animals 176
invasive plants 176–80
alien 178
land use practices 175
plant conservation programme 186–7
Gampaha Botanic Gardens 128
Garnotia seychellensis 199
Gastonia lionnetii 196
Gastonia seychellarum 197
genes loss by genetic derivation 65
germplasm conservation 11–22
current situation 14–22
granite islands
endemic species 195–9, 207
extinct (apparently) species 207
indigenous species 199
native species 208
threatened flora 206
Grewia glandulosa 201
Grewia picta 201

Hakgala Botanic Gardens 128
hardwood trees, indigenous 182–3
Hederorkis seychellensis 198
Hibiscus spp. 101–2
Hyophorbe amaricaulis 100
Hypoxidia maheensis 199

Impatiens gordonii 196
in situ conservation 169–246
Seychelles and threatened flora 204–6

indigenous flora in Galápagos islands 171–3
indigenous hardwood trees in Galápagos islands 182–3
indigenous species in Seychelles 199–204
indigenous vegetation in Galápagos islands 173–4
indigenous/alien plants in Galápagos islands 169–89
international co-operation for training 47–8
international co-operation in tropics 45–55
international issues 11–76
international staff training scheme implementation 54–5
introduced mammals in Galápagos islands 177
invasive animals in Galápagos islands 176
invasive organisms and land use practices 180–1
invasive plants, alien in Galápagos islands 178
invasive plants in Galápagos islands 176–80
involuntary selection and gene losses prevention 65

Jardin Exotique (Fort Napoléon) 72–4
Jubaea chilensis 153–6
Jubaea spectabilis see Jubaea chilensis
Justicia congestiflora 202

Kandy district 130–4
  medicinal plant extraction pattern 136
  medicinal plants utilization 137–43
Kew Royal Botanic Gardens and orchid propagation 290–4

land use in Brazil 121–2
land use in Galápagos islands 175, 180–1
Lautembergia aff. *neraudiana* 194–5
legal measures for species protection 61–2
legal measures for wild species protection 62–3
liaison *see* twinning and liaison
Lipizauga Botanical Sanctuary 241–6
  accessions 243, 245
  development 242–3
  endangered species 245
  the future 246
  orchid house 244
  records 245
  situation 241–2
  staff 245
  vegetation 242

local economy and development in Galápagos islands 181–3
long-term conservation in habitats 311–14

Madagascar and botanic gardens development potential 331–2
Maillardia montana 203
Malagasy threatened endemics 331–2
Malaysia and rôle of fernarium 223–38
Malaysian Borneo *see* Sabah
Malesian region, orchid knowledge in 303–4
Malleastrum depauperatum 201
mammals, introduced in Galápagos islands 177
Mauritius/Rodrigues
  endangered plants 107–12, 113–14
  ex situ collections 95–105
    case histories 100–3
    important plants for cultivation 103–4
    plant propagation 98–9
    plant reintroduction 99–100
medicinal plants
  conservation in Sri Lanka 127–34
  cultivation in Peradeniya Royal Botanic Gardens 144–9
  extraction pattern in Kandy district 136
  utilization in Kandy district 137–43
Medusagyne oppositifolia 196
Miconia robinsoniana 174

National Park zones in Galápagos islands 181–2
native species
  Aldabra 207
  granite islands 208

Obetia aldabrensis 203
Oeoniella polystachys 200
on site training in tropics 52–3
orchid
  botanic gardens 267–73
  conservation 267–326
    Bogor Botanic Gardens 297–305
      future of garden and collection 304–5
      orchid collections 299, 302–3
      orchid knowledge in Malesian region 303–4
    Sabah 307–26
      long-term conservation in habitats 311–14
      orchid flora 310–11
      short-term conservation 314–16
      Tenom Orchid Centre 316–23

Tenom Orchid Centre (cont.)
hybrids, preservation of 322–3
orchid flora 318–20
programmes progress 323–5
wild orchid preservation 320–2
wild species,
collection/cultivation of 316–18
flora 310–11, 318–20
fruit (*Dendrobium secundum*) 284
house (Lipizauga Botanical Sanctuary) 244
hybrids, preservation of 322–3
knowledge in Malesian region 303–4
propagation 289–94
communication 294
Kew Royal Botanic Gardens 290–4
development of collection 290–1
endangered species 292–4
plants, large numbers of 291–2
research 275–86
Eria pubescens 280
orchid fruit (*Dendrobium secundum*) 284
Orchidaceae subfamilies 281
Orchidales 279
phylogenetic 278–80
pollination biology 280–2
taxonomic 277–8
vegetative and sexual reproduction 282–5
wild species 316–18
see also wild orchid
Orchidaceae subfamilies 281
Orchidales 279
orientation programme in Brazil 122
ornamental ferns 232
overseas territories and wild flora protection 59–75

Pandanus pyramidalis 102
Papua New Guinea *see* Lipizauga Botanical Sanctuary
Peponium sublitorale 202
Peradeniya Royal Botanical Gardens 128, 130, 133, 144–9
phylogenetic research in orchids 278–80
Piper sp. 200
Pisonia aculeata 203
Pisonia sechellarum 198
plants
colonized zones in Galápagos islands 183
conservation programme in Galápagos islands 186–7
extraction patterns in Sri Lanka 132
inventories in Sri Lanka 132–3
propagation 115–17
Mauritius/Rodrigues and *ex situ* collections 98–9
record systems, computerized 335–6
reintroduction in Mauritius/Rodrigues and *ex situ* collections 99–100
rôle in local economy and development of Galápagos islands 181–3
trade and CITES 249–64
Platylepsis sechellarum 198
pollination biology and orchid research 280–2
propagation in fernarium 230–2
propagation techniques in fernarium 237–8
Psathura sechellarum 197
Pseuderanthemum tunicatum 200
Psychotria pervillei 202
Psychotria silhouettae 197

Quindío 89–94
Calarcá project 92–4
history 90
institutions contributing 91–2
planning 90–1

Ramosmania (*Randia*) *heterophylla* 103
Rapanea seychellarum 197
rare/threatened ferns 223–38
reintroduction
guidelines 211–15
into the wild 209–15
restoration biology 210–11
programmes 66–8
Ruizia cordata 217–22
research in ethnobotanical garden and tropical plant conservation 80–1
restoration biology and reintroduction into the wild 210–11
Rio de Janeiro botanical garden and Atlantic Forest conservation 163–6
Rodrigues *see* Mauritius/Rodrigues
Rothmannia annae 197
Ruellia monanthos 203
Ruizia cordata 217–22
male/female specimens 218
reintroduction 220–2
young plants 220

Sabah, orchid conservation in 307–26
Salvadora angustifolia 202
Scalesia incisa 172
scientific programme in fernarium 227–8
Second International Botanic Gardens Conservation Congress, resolutions

passed at 343–70
English 343–50
French 351–60
Spanish 361–70
sexual reproduction *see* vegetative and sexual reproduction
Seychellaria thomassetii 199
Seychelles and threatened flora 193–208
Aldabra 206
endangered species 207–8
endemic species of granite islands 195–9
ex situ conservation 206
extinct species 194–5
granite islands 206
in situ conservation 204–6
indigenous species 199–204
short communications 331–9
short-term conservation in Sabah 314–16
Sideroxylon grandiflorum 101
Smythea lanceata 195
South America and botanic gardens 25–32
background 28–9
national and international networks 29–31
species population safeguarding 64
species protection, legal measures for 61–2
species samples collection 64
Sri Lanka and medicinal plants conservation 127–34
achievements/recommendations 134
botanic gardens 128–30
constraints 133
importance 130–2
plant extraction patterns 132
plant inventories 132–3
staff training scheme, international 54–5
staff training in tropics 45–55
student volunteers, rôle of 53
sub-tropical trees, rare or threatened 151–60

Tabernaemontana coffeoides 202
taxonomic research in orchids 277–8
Tectiphiala ferox 101
Tenom Orchid Centre 316–23
thermic zones and Andean flora 89–94
threatened species identification 60–1
Toxocarpus schimperianus 198
TRAFFIC-France and tropical plant species traffic 259–64
means of action 261–3
training
course structures and curricula 53–4
course for tropics staff training 50–2
international co-operation 47–8
on site in tropics 52–3
tropical plant conservation *see* ethnobotanical garden and tropical plant conservation
tropical plant species traffic 259–64
tropical trees, rare or threatened 151–60
tropics, botanic gardens in 333–4
tropics and staff training 45–55
development 46–7
financial considerations and training courses 50–2
international co-operation for training 47–8
international staff training scheme implementation 54–5
rôle of Botanic Gardens Conservation Secretariat 49–50
on site training 52–3
student volunteers, rôle of 53
training course structures and curricula 53–4
twinning and liason in botanic gardens in tropics 333–4
Tylophora coriacea 202

university botanic garden development in Brazil 119–24

vanishing forests in Malaysia 224–6
Vateriopsis seychellarum 196
vegetation, indigenous 173–4
vegetation in Lipizauga Botanical Sanctuary 242
vegetative and sexual reproduction 282–5
Vernonia seychellensis 194
Viscum triflorum 203

watersheds in Galápagos islands 183–4
West Indies and WWF-France 71–5
Wielandia elegans 203
wild flora protection 59–75
wild orchid preservation 320–2
wild *see* reintroduction into
wild species protection, legal measures for 62–3
WWF-France and West Indies 71–5